EAST ASIA: ⌐
POLITICS, SOCIOLOGɪ,
CULTURE

Edited by
Edward Beauchamp
University of Hawaii

A ROUTLEDGE SERIES

South-South Transfer
A Study of Sino-African Exchanges

Sandra Gillespie

Routledge
Taylor & Francis Group
LONDON AND NEW YORK

First published 2001 by Routledge

2 Park Square, Milton Park, Abingdon, Oxon OX14 4RN
711 Third Avenue, New York, NY 10017, USA

Routledge is an imprint of the Taylor & Francis Group, an informa business

First issued in paperback 2016

Library of Congress Cataloging-in-Publication Data is available from the Library of Congress.

ISBN 978-0-8153-3870-3 (hbk)
ISBN 978-1-138-98266-6 (pbk)

Preface

This study aims to contribute to a larger understanding of international academic relations by investigating the transfer of knowledge on a South-South dimension. Specifically, it considers China's educational exchange programs for Africa, with a special concern for issues of sustainable and equitable development. A detailed description of the African student experience in China is provided through quantified data, obtained from a student questionnaire. Qualitative data, gathered through individual and group interviews with both African and Chinese parties, adds interpretive depth to the description.

This data is considered in the light of two bodies of literature related to views of international academic relations. A review of the literature on China's changing world view, with a focus on China's Third World Policy under the reign of Mao Zedong, provides the historical context for this study. The literature on the changing approaches to international academic relations, with a focus on the theories of the World Order Models Project, provides the theoretical context. These two transformative approaches to international relations form an evaluative framework in which Sino-African relations are examined within a global milieu. This study reveals that while Sino-African exchanges currently play a supportive role in the international arena, potential signs and strategies to promote greater equality are beginning to emerge from this South-South collaboration.

Acknowledgments

This study was made possible by the collaboration of many individuals and institutions to whom I would like to extend my sincere thanks and appreciation. The University of Toronto provided me with generous financial support in the form of a Graduate Fellowship, a Higher Education Award, and an Advance Planning Grant (UTAPS). The China/Canada Scholars Exchange Program contributed additional support that enabled me to travel to China to carry out my research from Tongji University in Shanghai.

While at Tongji, Mr. Yu Zeming, Officer for International Students, and Mr. Xu Kangnian, Director of the International Students Office, acted as my hosts and academic advisors. Mr. Yu and Mr. Xu arranged many of my appointments and provided the necessary guidance for me to conduct this research within China's institutes of higher education.

The African and Chinese participants within these institutes provided the initial inspiration and ultimate sources of knowledge for this study. I am particularly indebted to Messieurs Boniface Habonayo, Claver Ndayiragije, Cargele Nduwamungu, Evariste Nduwayo, and Charles Ntahomvukiye for garnering support and participation from the African communities across China. Special thanks to Evariste and Charles who accompanied me to all cities and sites that I visited. I would also like to acknowledge the extensive contributions of the African diplomatic community and in particular Ben Balthazar Rutsinga. Two other members played a significant role but felt it best to be unnamed.

Also remaining unnamed are many participants among the Chinese community. From among those that I can name, I would like to thank Yu Liming and Pei Chao. These colleagues supported me in undertaking this project and guided me in creating a protocol and set of instruments appropriate to the Chinese context. I was also fortunate to receive support and guidance from a strong academic committee, comprising of Drs. Ruth Hayhoe, Glen A. Jones, and George Sefa Dei.

Dr. Hayhoe provided a model of professional and intellectual supervision. She offered detailed and timely comments on every chapter of every draft, encouraging me to make use of the Galtung-Mazrui theoretical framework, modified by the specifics of China's historical background, to draw conclusions about Sino-African exchanges within the wider context of Comparative Education.

Dr. Jones also challenged me to reach further in my work. He was particularly helpful in the area of methodology, assisting me with the questionnaire design, statistical processing, and data analysis. Dr. Jones further contributed to many qualitative aspects of the project, helping me to clarify and articulate my personal location within this study.

Dr. Dei, through his detailed written response, also helped me to place myself within this project. He especially encouraged me to pursue an overall more critical edge to the work. His specific suggestions to critique Mao's and Galtung's theories and to consider the broader implications for pedagogy and global equity issues pushed me to develop what I hope is a more forceful conclusion.

This study has also benefited much from the critical input of a fourth reader, Dr. Julia Pan and the external examiner, Dr. Bernard Hung-kay Luk of York University. Dr. Pan posed insightful and thoughtful questions during the mock and final examinations and dealt with many sensitive aspects of this project in a compassionate and constructive manner. Dr. Luk's lengthy, highly detailed written report challenged me to refine many areas of this work. His specific suggestions for the historical chapter enabled me to clarify many crucial points. I thank him for his significant contributions and sincere interest in this project.

I would also like to extend my thanks to the late Dr. Geoff Isherwood of McGill University. Dr. Isherwood's unbounded enthusiasm for this subject motivated me to continue my studies and pursue this topic at the doctoral level. I appreciate the pivotal role he played.

Many others have helped along the way. Marg Brennan, Mary Macdonell, and Janice Verner provided courteous and efficient administrative services at every turn. Carolyn Coté, Senior Coordinator of the University of Toronto School of Continuing Studies Intensive English as a Second Language Program, always granted me maximum flexibility with my work schedule. Caroline Archambault read the entire final draft and offered valuable and incisive editorial suggestions. Caroline, along with Pei Chao, Bertram Etoundi, Yu Liming, Evariste Nduwayo, Charles Ntahomvukiye, Zhang Xiaoman, Yu Zeming, and Barbatus Gatoto also provided vital translation services.

Barbatus Gatoto deserves special mention. Barbatus assisted me at every stage of the project, giving me the benefit of his knowledge and experience of the five years he spent as an undergraduate in China. I deeply appreciate his insights, assistance, and friendship.

Finally, I have saved the most fundamental debts to the last. I would like to thank my family, especially my sister Mary Ellen, my Grandmother the late Stella Kelley, the deSouzas, the Archers, and my Grandmother Isabella Gillespie. Above all, I would like to thank my Mother Ellen Mary and my Father Alexander James for their support and encouragement which have sustained me throughout these many long years of my academic pursuits. I am truly grateful.

Contents

LIST OF CHARTS

LIST OF TABLES

CHAPTER ONE

Introduction

CONTEXT

Since its establishment in 1949, the People's Republic of China began to provide African students the opportunity for higher education. The life these students have typically encountered on a Chinese university campus has some rather striking characteristics. First, a complete and total society exists within the parameters of the university grounds. A wall defines these parameters and surrounds this entire society. All staff and students live within this wall. Two smaller walls stand inside the main wall; these walls surround the foreigners. The first wall surrounds the foreign "experts" (mainly Western professionals); the second wall surrounds the foreign students (mainly African men). These walls divide the society twice. The walls separate the Chinese from the non-Chinese and the foreign experts from the foreign students. Thus, three distinct societies live inside the school grounds: the Chinese, the Westerners, and the Africans. In a way, each group plays a different role in relation to the other. The Westerners, as foreign experts, teach local Chinese students; the Chinese professors, as local experts, teach the foreign African students. In the first case, the more commonly discussed 'First World' to 'Third World' (North-South) educational transfer occurs;[1] in the second case, the lesser known phenomenon of Third World to Third World (South-South) exchange takes place. This study explores South-South aspects of knowledge transfer by examining the lives of African students in China.

PURPOSE

The purpose of this investigation is to acknowledge the sustained educational cooperation between China and Africa and place this knowledge within a larger literature on approaches to international and academic relations. Within the context of international academic relations, knowledge, resources, and people have moved among nations for centuries (McMahon, 1988). Despite this long historical precedent, the present movement differs from the past in that it flows in a largely South to North direction (Altbach, Kelly, & Lulat, 1985). Today international students tend to migrate from the peripheries (the less industrialized Third World) to the centres (the major industrialized nations). Past research has reflected this phenomenon (Altbach et al., 1985).

Traditional research efforts to conceptualize and examine international study, identified by S. Spaulding and M. Flack in 1976 and by Y. G-M. Lulat in 1984, continue to dominate the literature. These efforts have focused on two principal sets of issues: sociopsychological consequences and cultural adaptation (Lulat, 1984: 207). Only very recently has this focus shifted to examining patterns of international study within a larger context of global political and economic relationships, as exemplified in the works of Cummings 1993, Goodwin 1993, McMahon 1992, Scott 1993, and Sutton 1993 (as cited by Barnett & Wu, 1995: 354-355).

Despite an emerging global view, research on international study continues to be conducted almost entirely from the perspective of the Third World student in the First World, to the exclusion of other perspectives. As a result, there continues to be a paucity of literature on the flow of international students among Third World nations themselves (Altbach et al., 1985; Lulat, 1984) This study aims to begin to fill this gap.

Since Emmanuel John Hevi's 1963 publication of *An African Student in China*,[1] no other intensive work, to my knowledge, has focused exclusively on this topic. In this text, Hevi relates his journey which began in November 1960 as he embarked from Ghana to China in the hopes of studying medicine. Hevi chronicles his personal experience and the experiences of 118 other Africans who arrived in Peking in 1961-62. This study attempts to complement Hevi's work and contribute to a larger understanding of international academic relations by directing attention toward a South-South dimension of knowledge transfer.

SIGNIFICANCE

Ideally, the findings of this research will contribute on both a practical and theoretical level to all parties concerned. The African and Chinese participants have both expressed hope that their sustained educational efforts may be further enhanced and that their experiences may be heard by a larger audience. For a Western audience, hearing about Sino-African experiences may broaden our knowledge about the different kinds of international educational transfers and may provide a context for understanding Western attitudes and aspirations in a global milieu. When, for example, Canadian international agencies set out to work with the Third World, it would be instructive to know how nations of the Third World work with each other.

Snow (1988: xiii-xvi) points out that often researchers in the West, pre-occupied by the complexities of their own relations with Asia and Africa, have taken comparatively little time to examine how the peoples of those regions have related among themselves. Indeed, the action and reaction of Western and non-Western values is a major theme of the modern world. Since Vasco daGama arrived in the Indian ocean at the end of the 15th century, the story of humankind has been largely an account of the response of Asia, Africa, and South America to the alien culture of the West (Snow, 1988: xiii). However, long before the West rose into prominence, contacts between other cultures flourished.[3]

Building international appreciation requires understanding ourselves, others, and how we relate together. It also involves understanding how others relate among themselves (Snow, 1988). In efforts to internationalize and build a truly global future, the consideration of contacts among all parts of the world becomes critical. The sustained cooperation in educational exchange which has taken place in the last fifty years between China and African nations may be an instructive example. This major phenomenon, which deserves more attention than it has received, is the focus of this study.

METHODOLOGY

In order to examine Sino-African educational exchanges, I supplemented quantifiable data, obtained from surveys of African students in China, with more qualitative information gathered through individual and group interviews with both African and Chinese parties. For the Chinese parties, I visited a total of six sites in two cities. I held seven interviews with a total of six participants. I spoke with professors, administrators, and government officials involved in these international academic exchange programs. For the African parties, I visited fourteen different sites in four cities. I held seven interviews with twelve participants from six different countries and collected survey data from 133 students from twenty-nine countries. In all, I spoke with current undergraduate and post graduate scholars, graduates living in China, and three Embassy Counsellors. In addition, I obtained institutional documents from three sources, including the State Education Commission. The specifics of the data collection are detailed in Chapter Four.

LIMITATIONS

I make no claim that these parties or sites provide typical representations of the experiences of African students in China. Instead, these sources illuminate a broad range of experiences, institutional settings, and educational environments which I hope will provide a multidimensional understanding to the study at hand.

As such, this research may not be easily generalized or replicated and thus may be seen to be limited in both external and internal reliability. However, issues of reliability to secure replicability are not an integral part of this investigative paradigm. Instead, this study is informed by the belief that the same or similar settings can mean different things to different people and even to the same person at different times (Borman, Goetz, & Le Compte, 1986; Hammersley & Atkinson, 1983).

Central to this understanding is the reflexive character of social research; that is, the recognition that both the researcher and the researched are engaged in the construction of knowledge (Borman et al., 1986; Davies, 1982; Eisner, 1991; Hammersley & Atkinson, 1983; Joron, 1992). In other words, the researcher, as the instrument that engages the situation and makes sense of it, is not detached from the process but rather an integral part of it (Eisner, 1991). Thus, rather than engaging in attempts to eliminate the effects of the researcher, this principle grounds personal experience as a starting point (Hammersley & Atkinson, 1983). An experientially grounded approach locates the researcher in the research process by recognizing that every way of knowing, every form of knowledge is organized from somewhere, a null point, a set of coordinates that mark the standpoint of the knower (Schutz, 1962, as cited in Jackson, 1991: 127; Joron, 1992).

One's "set of coordinates" may be partly defined by demographic characteristics, such as class, race, gender, culture, language, religion, age, marital status, education, and so on. As an occidental woman from the Western world attempting to conduct research, primarily in English, about the experiences of a diverse African scholarly community in the People's Republic of China, my "way of knowing" was both necessarily informed and limited by my set of coordinates. Attempting to account for one's location in the process of trying to account for the perspective of others is one goal of the reflexive research process. bell hooks (1988) conditionally supports such endeavours:

> Certainly it is important and necessary for people from any ethnic/racial group to play a significant role in the creation and dissemination of material about their particular experience. It is equally important for all of us to work at learning more about one another, and such learning is often best expressed in concentrated work and study on another group. (hooks, 1988: 46)

While hooks (1988: 42-48) encourages concentrated work and study to learn about one another, she cautions that when we write about cultures or experiences of ethnic groups different from our own, the issue of who will be regarded as the "authoritative" voice is likely to become political. hooks stresses the necessity to "actively refuse" a position of authority. One way of doing so is for researchers to attempt to locate themselves: to talk about their possible set of coordinates, their personal location to a given situation. I have begun to do this here and continue to do so in Chapter Four.

OUTLINE

This study is anchored in two bodies of literature related to views of international relations. In Chapter Two, the literature on China's changing world view, informed by the three main foreign policies of the Maoist reign, the 'Peaceful Coexistence' approach of the 1950s, the 'Revolution' approach of the 1960s, and the 'Grand Alliance' approach of the 1970s, provides the historical foundation for this study. In Chapter Three, the literature on the changing approaches to international academic relations, specifically, the theories of the World Order Models Project, informed by the four principles of Johan Galtung's model of positive action (equity, autonomy, solidarity, and participation) and Ali A. Mazrui's three strategies of African modernization (domestication, diversification, and counter penetration) provide the theoretical context. In Chapter Four, I outline the intended research design, discuss changes made, and detail the actual protocol that enabled me to collect data. In Chapters Five through Eleven, I present the findings of the study. The findings are organized into the following seven chapters according to the seven sections of the questionnaire: Student Profile, Motivation, Issues, Social Contact, Academic Experience, Chinese Language and Progress, and Financial Support. Within each chapter, I use the quantifiable data, obtained from the questionnaire, as a framework for the more qualitative information, obtained through the interviews with African and Chinese participants. In the concluding chapter, I return to the two approaches to international relations and attempt to place the findings within these views. Maoist transformative policies of international relations linked to transformative theories of international academic relations form an evaluative framework in which to examine and reflect upon the place and possibilities of Sino-African exchanges within a global milieu.

Notes

.[1] I use the terms 'First World' and 'Third World' as defined by Mao Zedong in his Three-Worlds Theory. This theory, first discussed by Mao in his interview with Zambian President Dr. K. D. Kaunda and later officially pronounced by Deng Xiaoping in 1974, is further explored in Chapter Two.

.² *The Dragon's Embrace* (1967) is Hevi's subsequent text related to the activities of China in Africa.

.³ In fact, seventy-five years before Vasco daGama sailed round the Cape of Good Hope to establish a Portuguese empire, the court of the Ming dynasty repeatedly sent great expeditions across the Indian Ocean to the ports of the East African coast (Snow, 1988). For a closer look at the nature and depth of these historical linkages, see *A Sketch of Sino-African Relations in Historical Perspective*, Appendix A.

CHAPTER TWO

China's Changing World View:
The Evolution of Foreign Policy during the Maoist Reign

INTRODUCTION

The findings of this study are largely informed by the voices of African students in China. However, an African perspective of Sino-African relations is not the initial point of entry but rather the other way around. As Ogunsanwo (1974:1) explains, Sino-African relations have largely resulted from the diplomatic initiatives of the People's Republic of China rather than those of African nations. This immediately begs a number of questions. Why is China interested in Africa? How does Africa fit into China's image of the world?

Samuel Kim (1979, 1980) proposes that China's image of world order is a corollary of its image of internal order and thus a projection of self image. China's behaviour in the international community can therefore be viewed as a reflection of its world image and self image. In this light, Kim (1979, 1980) advances the notion that these images integrate both normative and epistemological principles. On the one hand, these images embody dominant social norms and values. As such, they serve as philosophical assumptions about the international order. On the other hand, these images provide an epistemological paradigm. This paradigm performs cognitive, evaluative, and prescriptive functions; it leads policy makers to define the state of the world, to evaluate the meaning of the world, and to prescribe the correct behaviour to heed (Kim, 1979: 49, 1980: 16).

The theme that correct behaviour is a manifestation of correct thought permeates all important theoretical writings in the People's Republic of China. Such a notion is referred to as the Chinese 'world outlook.' Since 1949, China's world outlook has been largely shaped by Mao Zedong's thought. Therefore, China's definition of its place in the world during the Maoist reign serves as a useful context in which to explore China's global policies. Moreover, the Maoist image of world order provides an indispensable frame of reference for assessing any change or continuity in the post-Mao global policy (Kim, 1979, 1980).

PURPOSE AND FOCUS

The purpose of this chapter is to discuss the historical context in which African students first made their way to study in China. To do this, I trace the evolution of China's definition of its place in the world as reflected by the evolution of China's three main foreign policy strategies: the 'Peaceful Coexistence' strategy of the 1950s, the 'Revolution' approach of the 1960s, and the 'Grand Alliance' tactics of the 1970s (Lin, 1989; Yahuda, 1978, 1983). Within this purview, I examine components of the Maoist world vision and highlight China's policy towards the Third World. Finally, I point out that while all three strategies failed to survive in totality, each, in part, continues to influence current policies as China continues to define itself and its place in the world (Kim, 1980, 1983; Lin, 1989; Yahuda, 1978, 1983).

PHASE ONE: THE PEACEFUL COEXISTENCE APPROACH

The Peaceful Coexistence approach of the early 1950s had its intellectual roots in the 1940s. During World War II, Nationalist and Communist Chinese leaders, engaged in civil war, sought support from the emerging superpowers, the United States and the Soviet Union, respectively. When the Communists won victory in October 1949, they strengthened their alliance with the Soviet Union, the 'motherland of socialism', and began to share the Soviet view of the United States as the major imperialist adversary (Levine, 1989; Lin, 1989).

Two-World Theory

Just months before the official establishment of the People's Republic of China, Mao declared this alliance:

> Internationally we belong to the anti-imperialist front, headed by the Soviet Union. . . . The Chinese people must either incline towards the side of imperialism or that of socialism. There can be no exception to the rule. It is impossible to sit on the fence. There is no third road. . . .
> (Mao, *Selected Works*, IV, p. 415, as cited by Ogunsanwo, 1974: 3)

Articulating the principal contradiction of the postwar international system, Mao thus obliged all Chinese to lean in the direction of a socialist alliance with the Soviet Union (Kim, 1980, 1984). Mao looked to the Soviet Union as a model to emulate and claimed, "The Communist Party of the Soviet Union is our best teacher and we must learn from it" (as cited by Yahuda, 1978: 46).

However, Chinese and Soviet expectations of each other within this teacher/pupil alliance soon proved to be unrealistic. As the alliance began to falter, Mao continued the call to learn from the Soviet Union but began to stress the necessity for China to acquire an independent outlook:

> We must not eat pre-cooked food. If we do we shall be defeated. We must clarify this point with our Soviet comrades. We have learned from the Soviet Union in the past, we are still learning today, and we shall still learn in the future. Nevertheless our study must be combined with our own concrete conditions. We must say to them: We learn from you, from whom did you learn? Why cannot we create something of our own? (Schram, *Mao Tse-tung Unrehearsed*, p.129, as cited by Yahuda, 1978: 106)

Intermediate Zones

Indeed, Mao did create something unique. In the process of establishing independence from the Soviet Union, Mao modified the prescribed Two-World Theory and raised the notion of 'an intermediate zone.' Instead of predicting an impending confrontation between the United States and the Soviet Union, as many had when the Cold War began, Mao declared the international situation to be 'extremely favourable' (Kim, 1980; Lin, 1989). In his paper tiger thesis,[2] Mao minimized the strength of the U.S. and the dangers of a Soviet-U.S. war. Mao proclaimed that the true battlefield now lay, not between the two worlds, but rather in the vast zone that separated the two rivals: a zone that included many capitalist, colonial, and semi-colonial countries across Europe, Asia, and Africa (Kim, 1984; Lin, 1989; Yahuda, 1978). This intermediate zone became the new ally for the socialist camp because it served as a protective buffer, constituting, in Mao's own words, 'the rear areas of imperialism' (Kim, 1980: 30-31). Thus, the theory of the intermediate zone, comprising of what later was referred to as the Second and Third Worlds, reflected Mao's changing perceptions of the international environment.

Henceforth, Chinese leaders were nurtured with a tripartite perspective of international relations (Kim, 1984; Lin, 1989). According to Lin (1989: 29), Mao's emphasis on the existence and importance of a third force enabled China to develop its own identity and expand its own influence in international relations. China could now maintain its ideological commitment to the Soviet Union and at the same time seek relations with other nations with whom it shared a more common historical experience and international stature. Theoretically, China had defined an area that belonged to neither the Soviet Union nor the United States. This middle ground not only served China's own interests but the larger interests of the socialist world as well. Mao's assertion of a changing world order and China's place within it later crystallized into his Three-Worlds Theory. This notion of three worlds, though not yet fully developed, began to influence

China's conception of the world in general and the Third World in particular. In fact, China's first articulated Third World policy of Peaceful Coexistence was premised upon this tripartite perception (Lin, 1989: 229).

Five Principles of Peaceful Coexistence

The Five Principles of Peaceful Coexistence were first introduced by Zhou Enlai to an Indian delegate in 1953 and subsequently appeared in the Sino-Indian agreements on Tibet, signed in Peking on April 29, 1954 (Larkin, 1980; Lin, 1989). These agreements gave rise the following five principles:

> Mutual respect for each other's territorial integrity and sovereignty;
> Mutual non-aggression;
> Mutual non-interference in each other's internal affairs;
> Equality and mutual benefit;
> Peaceful co-existence. (Ogunsanwo, 1974: 7)

In essence, four of the five principles decreed a hands-off policy towards other sovereign states. The remaining stipulation, to seek equality and mutual benefit, was both a political and economic guideline. Thus, the five principles could be reduced to two: justice and non-interference (Larkin, 1980: 66). Though criticized as vague and platitudinous by some, as a doctrine these principles constituted a set of rules to govern international behaviour (Larkin, 1980; Lin, 1989). As a strategy, these principles revealed China's desire to create a united, self-conscious, anti-colonialist, and anti-imperialist coalition among newly independent countries. Such a broad coalition was to manoeuvre China out of isolation and secure its rightful position in the world. To this end, China tempered ideological differences and extended reconciliatory policy initiatives as a means of approaching many Third World countries (Lin, 1989: 230).

As early as August 11, 1954, the relevance of these principles extended beyond Asia, as Zhou Enlai declared them the basis for 'relations between China and the various nations of Asia and the world' (Gittings, 1974, as cited in Larkin, 1980: 66). The following year, Zhou firmly established these principles as China's official state policy towards other Third World countries at the Bandung Conference (Kim, 1984; Lin, 1989).

Bandung Conference

The Bandung Conference, held in Indonesia, April 18-27, 1955, was conceived by the Colombo Powers of Burma, Ceylon, India, Pakistan, and Indonesia and consisted of twenty-nine Afro-Asian states. The conference was not initiated by China, nor did China take a part in its planning. In fact, China was not even envisaged as a participant in the original proposal (Larkin, 1971; Ogunsanwo, 1974). However, contrary to expectations, China was not in the periphery at Bandung; in fact, for the first time in modern history, China played an active role, as an acknowledged, independent power, shaping the pattern of world order (Ogunsanwo, 1974; Yahuda, 1978).

Bandung signified China's modern debut onto the world stage, and this debut marked a watershed in Chinese diplomacy (Cooley, 1965; Ogunsanwo, 1974). Zhou Enlai, representing the Communist delegations, strove to resolve outstanding differences and establish a reputation for reasonableness. He avoided conflict, sought reconciliation, and steadfastly identified China with the common cause (Ogunsanwo, 1974). He even managed to introduce two additional principles to the original five:

Respect for the freedom to choose a political and
economic system;
Mutually beneficial relations between nations.
(Cooley, 1965: 11)

However, Zhou's 'master card' was his offer to negotiate with the United States, China's main rival, on the Taiwan issue. Zhou made this proposal at precisely the right moment to achieve the desired effect (Larkin, 1971; Ogunsanwo, 1974). At the conference, Zhou gained prestige with a triumph of personal diplomacy, and China gained a reputation for being accommodating and ready to resolve differences by negotiations (Yahuda, 1978). In the wake of Bandung, moderation and neutralism emerged as positive forces in Chinese foreign policy (Kim, 1980; Lin, 1989).

In addition to marking China's diplomatic debut and new current of moderation, Bandung also marked the beginning of China's Third World dimension of foreign policy.[1] Zhou utilized the setting as a platform to establish China's Third World credentials, stressing two points that China shared in common with all the other countries: a history of colonial dominance and a need for further independence based on economic reconstruction (Yahuda, 1983; Lin, 1989). Significantly, neither of these points applied to the Soviet Union. From this time on, China's foreign policy drew away from the Soviet clutches and drew towards embracing a common identity with former colonial countries. Henceforth, the Chinese leadership attached increasing significance to Afro-Asia as the primary centre of the anti-imperialist struggle, and Afro-Asia solidarity, as embodied by the 'spirit of Bandung,' became a prominent theme in Chinese pronouncements (Yahuda, 1978, 1983).

The proposed image of Afro-Asia, that Asians and Africans share a common political and social task, provided powerful rhetoric. Bandung initiated the articulation of a Third World voice and this voice was to be heard in the global arena thereafter. In this way, Bandung was of great and lasting symbolic significance. Beyond symbolism, however, the spirit of Bandung soon diminished.

Strategy Downfall

The spirit of Bandung never materialized into broadly effective institutions nor did it create any substantial mechanisms for ongoing relations. China was unable to harness the momentum gathered at Bandung and unable to establish any type of extensive relations with the Third World. Ultimately, the strategy failed to create a viable foreign policy framework. Lin (1989: 231) explains that both domestic and international factors contributed to this failure.

Domestically, the increasing radicalization within China, as evidenced by movements such as the Anti-Rightist Campaign (1957) and the Great Leap Forward (1958-60), made a moderate foreign policy line politically unappealing. Internationally, the second Taiwan Strait crisis in 1958 and the subsequent efforts of the United States to contain China diminished hopes of maintaining a stable environment. China's calls for the unity of the Third World were further muted by the looming Sino-Soviet split, as China now found it increasingly necessary to distinguish pro-Soviet from pro-Chinese countries. Moreover, China's relations with neighbouring nations became increasingly strained as outstanding boundary and territorial problems emerged. According to Yahuda (1996: 55), China's neighbours feared that a newly reunified China would be influenced by the legacy of its imperial past that wielded a superior lordship over other Asian rulers. In addition, Beijing's commitment to communism deepened their distrust. Beijing was perceived as a supporter of local communist parties dedicated to the overthrow of the newly established and fragile regimes. These neighbours feared that China would exploit domestic weaknesses as well as inter-regional disputes. In 1958, when Chinese foreign policy shifted away from the moderation of Bandung towards a more militant revolutionary line, these misgivings about China intensified (Yahuda, 1996: 55).

PHASE TWO: THE DOMINANCE OF REVOLUTION

In the late 1950s and early 1960s, both domestic and international conditions created an atmosphere in China in which the low-key, conciliatory approach of Peaceful Coexistence was replaced by a revolutionary-based strategy towards the Third World. Two developments that especially contributed to this revolutionary spirit in China were the escalating tensions between China and the Soviet Union and the growing independence movements in Africa (Lin, 1989).

The Collapse of the Sino-Soviet Alliance

Since the establishment of the People's Republic of China, the Soviet factor had been at the centre of Chinese politics (Yahuda, 1978: 102-103). China's ideology, economy, national security, and foreign policy were all based on the 'leaning to the side of' the Soviet Union in a bipolar world. The rupture with the Soviet Union fundamentally altered this paradigm. This rupture did not emerge suddenly but rather unfolded in the process of the deteriorating Sino-Soviet alliance. The alliance, seemingly cemented by the Korean War, began to unravel as historical, cultural, and socioeconomic differences surfaced and proved to be irreconcilable. Eventually, differences over international politics and strategy drove the ultimate wedge between China and the Soviet Union (Yahuda, 1996).

Yahuda (1996: 57-58) explains that the Soviet Union, as the senior partner, could not permit China to jeopardize the Soviet's global interests. Concurrently, an independent China could not permit itself to be made subordinate to the Soviet Union. These tensions affected the very nature of the entire international communist movement. Moreover, because ideology was at the core of the legitimacy of the movement, these differences were expressed in ideological terms. Therefore, the legitimacy of each regime was challenged by these rising disputes. Ultimately for Marxist-Leninists theorists, only one correct view could exist and no true comrade would persist in publicly putting forward a contrary perspective. By the early 1960s, both sets of leaders accused the other of betraying the communist cause and their own people (Yahuda, 1996: 58).

One root of this complex problem stemmed from their respective relationships with the United States. After Stalin, Khrushchev sought to ease tensions with the USA, in part to carry out reforms at home and to reduce the costs and risks of maintaining nuclear weapons. At the same time, the Chinese leaders also sought to diffuse tensions with America, in order to focus on domestic economic development. Unlike their Soviet colleagues, however, Mao Zedong and Zhou Enlai found the Eisenhower administration unresponsive. American policy held that the way to divide the two communist giants was to keep up the pressure on China; thus, they denied to China the diplomatic overtures extended to the Soviets. As a result, China found no reason to believe Khrushchev's claims that the United States had moderated its tactics (Yahuda, 1996: 58-59).

These differences quickly escalated into issues of national and international security (Lin, 1989). In 1959, when the Soviet Union and the United States joined forces to restrain China from developing nuclear arms, it was a point of no return.[4] In the following year, the Soviets dealt the Chinese leaders a huge blow by withdrawing all economic aid and technical expertise (Yahuda, 1996). In 1962, China, still reeling from this withdrawal and from the failure of the Great Leap Forward, faced three major crises at its borders. In Xinjiang, tens of thousands of Uigurs crossed into the Soviet Union; in the southeast, Taiwan posed the threat of an invasion; and in the southwest, a border war erupted with India (Yahuda, 1983: 35). In this war, the Soviet Union sided with Britain and the United States in support of India and thus confirmed China's worst fear: an unholy triple alliance between the reactionaries (India's ruling class), the revisionists (the Soviet Union), and the imperialists (the United States) (Yahuda, 1983: 35). From this point forward, Chinese and Soviet leaders took opposite positions on all key international issues (Yahuda, 1996: 59).

Impact of the Collapse

The impact of the Sino-Soviet collapse on international politics in the Asia-Pacific region was not immediately obvious. By contrast, the impact on the Third World was keenly and immediately felt. After the split, both the Soviet Union and China increased their efforts to enlarge their own geographical stake on the international scene (Lin, 1989). From the early 1960s, the two nations competed for the allegiance of the various liberation movements and newly independent countries in the Third World (Yahuda, 1983).

Chinese leaders felt that their Third World policy had to be more militant in support of their strong criticism of the Soviet's alleged 'sell out' by the 'revisionist' Khrushchev. Thus, partly to respond to the split and partly to challenge Soviet dominance, China adopted a distinctively radical, revolutionary-based Third World policy throughout the rest of the sixties (Lin, 1989).

Independence Movements in Africa

Given their historical, demographic, and geographical ties, South East Asia was a primary concern for China's Third World policy. In many of these Asian countries, their independence was accompanied by well-developed social and economic infrastructures. By contrast, many newly independent African states were open to new social and economic models. Thus, African decolonization contributed to China's revolutionary zeal in that it provided a rare opportunity for China to put its new revolutionary-based policy into practice.

Chinese leaders felt that the modern Chinese revolutionary experience provided them with the insight to understand and deal with the problems of the African continent. Moreover, they felt that knowledge of this Chinese revolution could help Africans deal with African problems. At one point, Chinese leaders actually proposed to teach Africans Chinese history so that the Africans might better understand *African* conditions:

Africa itself looks like the seven powers of [China's] Warring States [403 BC to 221 BC] with its Nasser, Nkrumah, Hussein [sic], Sekou Touré, Bourguiba and Abbas [sic], each with his own way of leading others. In general everyone is trying to sell his own goods. Africa is now like a huge political exhibition, where a hundred flowers are truly blooming, waiting there for anybody to pick. But everything must go through the experience of facts. History and realistic life can help the Africans to take the road of healthy development. We must tell them the Chinese revolutionary experience in order to reveal the true nature of both new and old colonialism. In Africa we do no harm to anyone, we introduce no illusions, for all we say is true. (Cheng, The Politics of the Chinese Red Army, p. 484, as cited by Yahuda, 1978: 125)

Thus, the Chinese historical experience was advanced as a useful framework within which African conditions could best be understood. China's attempt to apply and universalize its experience to Africa was further revealed by Foreign Ministers Chen Yi's remark 'our yesterday is their today and our today is their tomorrow' (as cited by Yahuda, 1978: 125). Thus, attracted by a perceived common past and ripe revolutionary opportunities, China attempted to implement its new Third World policy in Africa.

Two Components of the Revolution Approach

Unlike the Peaceful Coexistence approach, which was vague and fragmented, the Revolution approach was comparatively concrete and systematic. The Revolution strategy had two major components. First, China supported countries fighting for independence or struggling against reactionary regimes. Second, China advocated self-reliance (Lin, 1989: 232-233).

Component One: Symbolic and Substantive Support

China increased both symbolic and substantive support for African countries undertaking various forms of struggle. In particular, Chairman Mao Zedong, on behalf of 650 million Chinese people, declared "full sympathy and support for the heroic struggle of the African people against imperialism and colonialism, . . . [and he expressed] firm confidence that ultimate victory would certainly be won" (Chinese African-People's Friendship Association, 1961: 3).

To promote the common struggle, the Chinese Communist Party reached out not only to national leaders but also to ordinary citizens. As early as 1949, China began offering Africans opportunities for higher education (see Appendix L). Bringing with them an image of an altruistic Communist government, young African scholars thus arrived in Beijing, (Sullivan, 1994). Upon arrival, these African students were welcomed with extraordinary fanfare:

> They were carried shoulder-high, showered with flowers and confetti and bombarded with the din of traditional rejoicing, gongs and cymbals and fire-crackers. They were led before microphones to voice their demands for freedom to applauding crowds half a million strong. They were borne round in limousines like ministers and seated beside the Chinese leaders at rallies and parades. . . . Very humble Africans, unknown young men and women, were received with honour by the greatest personalities of the land. . . . [Many] . . . found themselves closeted, almost as a matter of course, with Mao, his Prime Minister Zhou Enlai, his Foreign Minister Chen Yi; or all of them. (Snow, 1988: 73)

Beyond the symbolic fanfare, however, China delivered substantial support for revolutionary activities. In a systematic study of China's support for wars of national liberation in 1965 (the peak year of the Revolution strategy), Peter Van Ness (1970) examined three questions which were critical to China's state policy: one, did the relevant state have diplomatic relations with Peking; two, did they vote in favour of admitting Peking to the UN in 1965; and three, did their trade relations with Peking exceed $75 million in 1964 and 1965 (as cited by Yahuda, 1978: 159). Van Ness tested whether the nature of state-to-state relations correlated better than officially articulated revolutionary theory. He concluded: "Whether a foreign non-Communist country was seen to be 'peace-loving' or ruled by 'reactionaries,' or whether a Communist Party state was viewed in Peking as 'socialist' or denounced as 'revisionist' largely depended on the extent to which that country's foreign policy coincided with China's own" (as cited by Yahuda, 1978: 159; Lin, 1989).[6] In fact, during this period, China endorsed revolutionary armed struggle "in only 23 of a possible total of some 120 developing countries" (Van Ness, 1970, as cited by Lin, 1989: 233). Lin (1989), however, firmly asserts that China would have supported more countries if it had been able.

Snow (1988: 144-85) concurs that China offered what economic assistance it could. In the following two decades, Africa had become the object of a philanthropic crusade. The Chinese government spent approximately US$2 billion in loans, food, and aid projects.[7] Regardless of China's own domestic problems, assistance to Africa was to be a heroic endeavour: the poor helping the poor (Snow, 1988: 145).

Component Two: Advocacy of Self-reliance

China's willingness to provide economic aid to other Third World countries, even though it was far from rich in resources itself, supported the prevailing view that the struggle for political independence would be incomplete unless followed by a nationalized, self-sufficient economy. To that end, China urged newly independent states to develop a strategy distinct from those of the Western imperialists or Soviet revisionists (Lin, 1989). Thus, in addition to lending symbolic and substantial support, the second component of the Revolution strategy involved advocating self-reliance. Shih (1993) explains:

> One of China's missions in the Third World is to help these nations achieve self-reliance in order to sever links with imperialism and facilitate its eventual collapse. Although China does not have the resources of a superpower, China can demonstrate its sincere support in every possible fashion and without political strings. In an anticolonial struggle, China will sometimes back all the factions involved. Third World nations are expected to appreciate truly friendly support and gradually phase out the politically motivated assistance given by other powers. This is probably why the Chinese deem South-South cooperation critical to overall development of the Third World. The notion of South-South cooperation extends the scope of self-reliance to include the Third World as a whole. Receiving aid from China is thus more desirable than receiving it from a non-Third World nation. The stress on self -reliance portrays China as a model and the Chinese presence as being morally appealing. (Shih, 1993: 175-176)

This message of self-reliance and self-sufficiency was brought directly to the African continent by Foreign Policy Minister Zhou Enlai during his 1963-1964 tour. The dominant theme of Zhou's visit, the call for a new, independent, and prosperous Africa, was warmly welcomed. Six more African countries established diplomatic relations with China in that year alone (Lin, 1989). Zhou's tour represented a breakthrough; never before had China been so positively received.

Closer to home, however, China's revolutionary initiatives were less favourably greeted. The major diplomatic victories that China did realize, such as the signing of border treaties with neighbouring states including Burma, Nepal, Pakistan, Afghanistan, and Mongolia, were overshadowed by failures. China's increasingly militant stance, particularly the encouragement of domestic revolutions, fuelled existing suspicions among its neighbours. Except for a few countries such as Indonesia and North Vietnam, most Asian nations responded to these hard-line policies with caution. Lin Piao's famous 1965 article on the universal application of 'People's War' confirmed these misgivings (Lin, 1989).

Failure of the Revolution Strategy

By the mid 1960s, the Revolution strategy began to unravel. In Southeast Asia, China lost one of its last remaining allies in the area when diplomatic ties with Indonesia were severed in response to the bloody coup of 1965 (Lin, 1989). Even in Africa, the initial enthusiasm for the revolutionary spirit had been replaced by a more sober appreciation of its limits. China's hopes of revitalizing the spirit at a second Asian-African conference were quashed when the conference was cancelled because of a coup in Algeria, the host country. In addition, a series of subsequent coups drove out many African leaders who had close ties with China. This led to the expulsion of many Chinese diplomats and contributed to growing suspicions about China's presence on the continent. In the end, not a single government or movement significantly expanded their power because of China's revolutionary tactics (Larkin, 1971, as cited by Yahuda, 1978: 158).

According to Lin (1989), the Revolution strategy failed for two main reasons: overreaching and miscalculation. First, by committing to a broadly defined goal of revolution, China overextended itself into too many regions. The intense ideological component of the revolutionary strategy inhibited China from establishing priorities and developing effective means to implement them. Second, by relying on an ideologically based strategy, China seriously miscalculated the complexity and diversity of the Third World. China alienated itself from many countries by insisting on a united Third World struggle against both revisionism and imperialism. China failed to consider that each country had its own conception of national interest and wanted to define its own relationship with the superpowers. Moreover, China misinterpreted international trends and its own ability to influence world events. Chinese leaders believed that the Revolution approach, like the Peaceful Coexistence approach, would encourage a movement that would inevitably lead to vast changes benefiting all Third World nations. When this did not happen, Chinese leaders were forced to reevaluate and reorient their foreign policy strategy (Lin, 1989: 234-235).

PHASE THREE: THE GRAND ALLIANCE

The transition from revolutionary chaos to pragmatic reconstruction began in late 1968 and culminated in April 1969 at the First Plenum of the Ninth Chinese Communist Party Congress. Based on the concept of a united front of China, the U.S., and sympathetic Third World countries against the Soviet Union, this strategy ushered in a new era of Chinese foreign policy (Kim, 1980). Once again, domestic and international pressures combined to prompt the changes. Domestically, the disruptive effects of the Cultural Revolution, which had put Chinese foreign policy in limbo between 1966 and 1968, were subsiding. As the frenzy waned, Mao and other top leaders shifted their focus to more threatening developments, especially the

Soviet invasion of Czechoslovakia in 1968 and the Soviet border clash with China in 1969. The risk of Soviet military intervention against China, rendered plausible by the upheaval from the Cultural Revolution, pushed Beijing to reassess its foreign policies (Levine, 1989; Lin, 1989). The reassessment followed a new analysis which identified four contradictions in the world, instead of the idealized one contradiction.[8] Kim (1980: 32-33) explains that Mao's difficulty in identifying the single principal contradiction revealed his 'agonizing reappraisal' of the international system. Kim (1980) adds that the structural shift from bipolarity to multipolarity, coupled with the Sino-Soviet split, prompted Mao to examine different variations on the theme of multiple zones.[9] Ultimately, because of the superpowers' hegemonic 'contention and collusion' in both intermediate zones, Mao proclaimed it desirable to combine the two zones in order to create the broadest united front (Kim, 1980: 32-33). The *People's Daily* newspaper was used to reinforce this united front stand and to argue that the immense changes of the late 1960s had led to this new historical situation:

> For a time US imperialism remained the arch enemy of the people of the world. But many countries in its camp were no longer taking their cue from it and most countries in Asia, Africa and Latin America won independence. Meanwhile the Soviet leadership betrayed socialism, restored capitalism at home and the Soviet Union degenerated into a social imperialist country. Then, after a succession of grave events, the Soviet Union not only turned into an imperialist superpower that threatened the world as the United States did, but also became the most dangerous source of another world war. (The People's Daily, as cited by Yahuda, 1978: 245)

Thus, the United States was portrayed to be on the defensive and in decline (largely as a result of the protracted Vietnam War) while the Soviet Union, as a younger imperialist power, was depicted as on the offensive and on a ruthless and insatiable incline (Yahuda, 1978). Henceforth, Soviet social-imperialism, rendered "more crazy, adventurist, and deceptive" than U.S. imperialism, became China's number one enemy (Kim, 1980: 33).

Alignment with the United States

The Soviet threat to Chinese security provided a rationale for establishing a temporary strategic alignment with the United States: the imperialist superpower that, though weakened, remained the sole power able to counter this danger (Levine, 1989). China's perceptions of the Soviet Union began to coincide with the United States' own anxiety over unprecedented Soviet expansion. Subsequent Nixon-Kissinger advances enabled China to shift its international strategy of opposing both superpowers to opposing only the Soviet Union. China then could embrace the United States as an implicit ally (Lin, 1989). Levine (1989) concludes that for both parties, a classical balance-of-power politics prevailed over ideology. The Sino-American rapprochement of the 1970s, culminating in the normalization of formal diplomatic relations on January 1, 1979, was rooted in a shared strategic assessment representing the union of two parallel obsessions: America's Cold War obsession with the Soviet Union and Maoist China's latter-day obsession with Soviet social-imperialism (Levine, 1989: 67-68).

The Sino-American rapprochement, the new Grand Alliance, held tremendous practical implications for China (Lin, 1989). First, the rapprochement enabled Beijing to establish new contacts with other industrialized nations. During the final years of Mao's rule, political and economic relations flourished with the West. In 1973, for example, China purchased US$4.3 billion worth of industrial equipment from the West, the largest such move made since China accepted Soviet aid to construct its industrial base in the 1950s (Lin, 1989). Second, as a result of improved relations, Beijing was able to forge new contacts with pro-U.S. developing countries. Between 1971 and 1972, twenty-four Third World countries opened or resumed diplomatic relations with China. In short, the reorientation of China's foreign policy in the early 1970s put China in a far better position to implement its Third World policy (Lin, 1989). After the initial focus on the Sino-U.S. rapprochement, Mao turned his attention to a more systematic and theoretical basis for this new arrangement (Kim, 1980).

Three-Worlds Theory

Mao's image of China and of the new world order finally crystallized in his Three-Worlds Theory (Kim, 1984: 183-184). The Three-Worlds Theory began to develop as a response to the increasing untenability of the 'lean-to-one-side' policy that Mao himself had earlier pronounced. As discussed, Mao premised that policy on a Two-World Theory which he later modified by the notion of intermediate zones, comprising what he later termed as the Second and Third Worlds. According to Kim (1984), Mao's repeated attempts in the 1960s to define the theory of intermediate zones in the face of a rapidly changing world revealed an acute crisis of Chinese identity. By the early 1970s, however, Mao finally resolved the crisis by positioning China with the Third World. Thus, within the final refinement of the theory of the intermediate zone emerged a model of the Three-Worlds.

The Three-Worlds Theory was officially pronounced by Deng Xiaoping at the Sixth Special Session of the U.N. General Assembly on April 10, 1974 (Kim, 1985; Yahuda, 1978). Three months earlier, however, Mao first discussed this theory in an interview with Zambian President Dr. K. D. Kaunda. At this meeting, Mao stated:

> In my view, the United States and the Soviet Union form the first world. Japan, Europe and Canada, the middle section, belong to the second world. We are the third world. . . . The third world has a huge population. With the exception of Japan, Asia belongs to the third world.

The whole of Africa belongs to the third world, and Latin America too. (Renmin Ribao (People's Daily), "Chairman Mao's Theory of the Differentiation of the Three Worlds Is a Major Contribution to Marxism-Leninism," translated in *Peking Review*, no. 45 (November 4, 1977), p. 11, as cited by Kim, 1980: 33)

Kim (1984) explains that the Three-Worlds Theory is a simplified model to define and assess the main contradictions in the international order. The theory operates as a geopolitical compass for China to establish its rightful place in the world. Like the Wallerstein world-system model,[10] which divides the global political economy into core, semiperiphery, and periphery, the Three-Worlds Theory also makes a tripartite division of the globe: the First World of two superpowers in predatory competition or collusion; the Third World of developing nations in Asia, Africa, and Latin America; the Second World of Northern developed countries in between (Kim, 1984: 183). Kim (1984) captures the essence of the theory in the following synopsis:

> Stripped to its core, Mao's Three-Worlds Theory is a theory of anti-hegemonism designed to strengthen the weak and the poor (including China) to overcome the strong and the rich. It envisions a united front strategy, derived from China's own revolutionary experience, that has been extrapolated to the global setting to pit the nations of the Third World against those of the First in an unfolding struggle to transform the postwar international system. Although the theory calls for a dual-adversary approach directed against both superpowers, in practice the Soviet Union has often been singled out as the greater threat to world peace. (Kim, 1984: 184)

The Three-Worlds Theory served different purposes at different times.[11] It initially served as a theoretical underpinning for the drastic shift in China's foreign policy. It also negated any unfavourable reactions some Third World nations felt in regard to the rapprochement and the subsequent close relations between China and the United States. Soon, however, the theory became a convenient tool to justify China's focus on the Soviet Union. Finally, and perhaps most importantly, the Three-Worlds Theory supported the Grand Alliance strategy with the United States which strongly influenced the development of China's Third World policy (Lin, 1989).

The Grand Alliance strategy affected China's Third World policy in two main ways. First, the strategy tied China's Third World policy closer to its concerns regarding the two superpowers. As China's relations with the two superpowers changed, so did its Third World policy. As a result, China's Third World policy became less coherent. Second, under the Grand Alliance strategy, China tended to judge other Third World nations according to their degree of 'Soviet connections' (Lin, 1989; Yahuda, 1978). Supporters of the Soviet Union were enemies and those who were not were allies.[12] Throughout the 1970s, China justified this stark delineation by two assertions: one, in the wake of the U.S. decline, the USSR was the most threatening hegemon; two, the Third World was the target of Soviet expansion. China thus urged all Third World countries to 'wake up' and support a broad coalition to contain the threat from the Soviet Union (Lin, 1989).

Demise of the Grand Alliance

Within a few years, three fundamental limitations of the Grand Alliance strategy began to surface. First, the strategy proved to be too simplistic. By insisting on the anti-Soviet criterion, many Third World states were alienated. By focussing largely on the single anti-Soviet factor, this strategy, like the Revolution approach, underestimated many developing nations' will to determine the nature of their own foreign relations. Second, the increasing parallels between Chinese and American policy on many Third World issues, especially those involving regional disputes, incited suspicions and resentment from many who viewed these parallels as evidence of China's increasing deviation from its proclaimed Third World position (Lin, 1989). However, as the strategy distanced China from many potential Third World allies, it furthered attached China to the United States. Thus, the third and perhaps the most serious flaw of the plan was the over-reliance on compatible and sustainable relations with the U.S. which, in the end, left China somewhat isolated. In the early days of the Reagan administration, any illusions of a Sino-American partnership were quickly exposed as the new American government took an increasingly pro-Taiwanese stance, accompanied by a revived U.S. Soviet rapprochement. While this new U.S. stance did not escalate into a major confrontation, it did signify the beginning of China's disenchantment with the Grand Alliance Approach (Lin, 1989).

Dropping the Grand Alliance

Above all else, the most important motive for dropping the Grand Alliance stemmed from China's increasing concentration on domestic reform and modernization (Lin, 1989). With the main benefit of the strategy, the normalization of diplomatic relations with the U.S. now exhausted, the continuation of the policy would only increase costs with no return. Moreover, to invest huge resources in direct conflict with the Soviet Union became counterproductive. To continue to distinguish between pro and anti-Soviet states only limited China from expanding relations. Moreover, harsh anti-Soviet propaganda now seemed outdated as the domestic scene de-radicalized (Lin, 1989). Indeed, internal affairs took precedent, particularly in the two years immediately following the deaths of Zhou Enlai and Mao Zedong in 1976. In fact, due to the serious domestic situation in the immediate post-Mao period, Beijing was in no position, militarily or otherwise, to employ provocative tactics. Thus, for a while China assumed a relatively passive position in which it remained before taking its first step in the post-Mao years of foreign policy (Sutter, 1986).

CONCLUSIONS

During the Maoist reign, China's definition of its place in the world underwent a protracted struggle (Kim, 1980: 34). The 1950s witnessed a dialogue between the two-camp theory and the theory of the intermediate zone, as China made its diplomatic debut with the Five Principles of Peaceful Coexistence at the Bandung conference. The spirit of Bandung, however, was short lived. As China continued to search for a place in the rapidly evolving international system, revolutionary tactics seemed more promising. The identity crisis of the 1960s, evident in China's repeated efforts to define the theory of the intermediate zone, was manifested in the break with the Soviet Union, in the conflicts with neighbouring nations, and in the loss of credibility on the African continent. Finally, however, China made peace with itself as a member of the Third World. China's foreign policy may thus be seen as an adjustment of struggles — of conflict, competition, coexistence, and cooperation — whose focus shifts from time to time, place to place, and actor to actor (Kim, 1980: 21, 34).

Notes

.¹ Kim (1980: 20-21) explains that the Law of Contradictions is a central idea in Mao's world view. Mao's philosophy of life revolved around the belief that contradiction is inherent in the social process itself and without it no social progress will occur. Every contradiction represents an objective reality. To resolve contradiction is to engage in a protracted struggle because as the moving force in nature, contradictions rise, resolve, and rise again. In the relationship between various contradictions, one and only one is the principal contradiction that necessarily determines the development of the others. In every given situation, the crucial task of leadership is to first identify and then resolve the principal contradiction. The remaining problems (the secondary contradictions) can easily be solved once subordinated to the resolution of the principal contradiction.

.² Mao launched his concept of imperialism as a "paper tiger" in an interview with Anna Louise Strong. Mao stated, "The atom bomb is a paper tiger used by the U.S. reactionaries to scare people. It looks terrible, but in fact isn't. . . . All reactionaries are paper tigers. In appearance, the reactionaries are terrifying, but in reality they are not so powerful. From a long range point of view, it is not the reactionaries but the people who are really powerful. . . . U.S. reactionaries, like all reactionaries in history, do not have much strength. . . . Although the Chinese people still face many difficulties and will long suffer hardships from the joint attacks of U.S. imperialism and the Chinese reactionaries, the day will come when these reactionaries are defeated and we are victorious. The reason is simply this: The reactionaries represent reaction, we represent progress" (Extracts from Mao's interview with Anna Louise Strong, August 1946, as cited by Hsuan Chi, IV: 1192-93 as cited by Schram, 1963: 279-280).

.[3] The Bandung conference marked the first opportunity for leaders of the six participating African states, Egypt, Ethiopia, Gold Coast (later Ghana), Liberia, Libya, and Sudan, to meet the rulers of Communist China (Ogunsanwo, 1974: 8). These meetings were followed by Chinese efforts to renew diplomatic, economic, and cultural contacts that had begun over 500 years ago. As Ogunsanwo (1974: 9) points out, in 1956, one year after the conference, Chinese cultural missions visited Egypt, the Sudan, Morocco, Tunisia, and Ethiopia. China also made commercial inroads into Africa with large cotton purchases from Egypt, followed by its first commercial contracts with other African countries, beginning with the Sudan and Morocco. On the diplomatic front, China succeeded in obtaining recognition from Egypt in May 1956 and the first Chinese embassy in Africa was established in Cairo. Ambassador Ch'en Chia-k'ang was sent to Cairo, where he remained until December 1965. From Cairo, China's diplomats began to follow the situation on the continent.

.[4] In 1959, the Soviet Union refused to supply China with a sample atomic bomb. A Test Ban treaty was signed in 1963. Undaunted, in 1964 China tested their first device (Yahuda, 1996: 59).

.[5] According to Yahuda (1978: 126), "it is critically important to note that China's didactic intentions did not call upon the Chinese to direct and instruct the Africans. Even though Peking was perceived as the true source of Marxism-Leninism and as the centre of principled opposition to imperialism, there was no sign that China's leaders sought to . . . direct its member constituents. . . . The Chinese position was still country-centred. . . . Mao had never practised the export of revolution."

.[6] Yahuda (1978: 159) cautions that perhaps Van Ness went too far in this conclusion. He believes that China's foreign policy concerns went beyond those of narrow state interest.

.[7] The Tanzanian-Zambian Railway, initiated in 1965 and competed in 1975 (two years ahead of schedule), stands as China's most significant accomplishment on the continent. The 'Tan-Zam' remains not only the longest railway in Africa (1,860 kilometres) but also the longest railway completed anywhere in the world since the end of the Second World War (Snow, 1988: 154).

.[8] According to Mao, the four major contradictions in the world now were as follows: one, oppressed nations versus imperialism and social-imperialism; two, the proletariat versus the bourgeoisie in the capitalist and revisionist countries; three, imperialist versus social-imperialist countries; and four, socialist countries versus imperialism and social-imperialism (Kim, 1980: 32).

.⁹ Mao's different variations on the theme of multiple zones included: one, the superpower zone, comprised of U.S. imperialism and Soviet social-imperialism; two, the socialist zone, made up of all Socialist countries; three, the first intermediate zone, including Asian, African, and Latin American; and four, the second intermediate zone, representing the major capitalist countries in the East and the West, except the two super powers (Kim, 1980: 33).

.¹⁰ The Wallerstein world system model is also applied to the analysis of international educational relationships, as discussed in Chapter Four.

.¹¹ While acknowledging the variety of purposes the Three-Worlds Theory served, Kim (1994: 129) critiques Mao's construction of Third Worldism and concludes it "was more symbolic than substantive." He states, "The cleavages in the South between fast-growing and stagnant, small and large, coastal and landlocked, left and right, and democratic and authoritarian made any claim of a unified Third World movement seem a curious mixture of rhetoric and wishful thinking. . . . The uneven and differentiated performance in economic growth and social equity has introduced a measure of distortion to the holistic image of the Third World."

While Kim interrogates Mao's use of Third Worldism, he also cautions against its quick dismissal. Kim (1994: 129) warns that "deconstructing the symbolism of the Third World as an independent force in world politics, if carried too far, can be just as misleading as the earlier claims on behalf of its negotiating solidarity. A more valid critique is normative and conceptual. The term "Third World" is increasingly challenged by those claiming to represent that world, who prefer such terms as "nonaligned" and "South" to a designation they see as unwittingly legitimating a hierarchy in the global political system. Without completely rejecting this critique . . . the label "Third World" [endures] partly because it persists in Chinese policy pronouncements and partly because it is emblematic of the common identity and shared aspiration that still link the countries and peoples of the poor South in an essential but elusive struggle to escape from poverty and underdevelopment."

.¹² Yahuda (1978: 159) points out that this way of thinking surfaced before. He states that there was "a natural tendency by the Chinese to perceive a linkage between opponents. During the Cultural Revolution their pronouncements often used the phrase 'anti Communist anti China' as if the one was equivalent to the other." Now the equation became 'pro-Soviet Union anti-China'.

CHAPTER THREE

Considerations of Theoretical Frameworks

INTRODUCTION

Perhaps the most significant Maoist legacy is the crystallization of the Three-Worlds Theory, linking symbolically and normatively China's fate with that of the Third World in the enfolding challenge to transform the existing world order (Kim, 1980: 34). To explore transformative possibilities of international order in relation to the flow of students from Africa to China, the field of Comparative Education provided a useful theoretical framework. Specifically, the field provided an arena in which to begin to understand the place of Sino-African educational cooperation within the international academic order.

Entering such an arena, however, was somewhat daunting. As Hans Weiler (1989) pointed out, in the field of Comparative Education consensus over epistemology has eroded. In other words, there is growing doubt over how we come to know what we know (see also Masemann, 1990: 471). What were the once seemingly fixed paradigms that have now apparently disintegrated?

PURPOSE AND FOCUS

The purpose of this chapter is to examine this question with the goal of understanding the broader underpinnings of my work and identifying the paradigm most appropriate to this project. To do this I trace, in a roughly chronological way, the changing approaches to Comparative Education. I begin by outlining the first three phases of methodological development in Comparative Education as delineated by Bereday (1964): the period of borrowing, prediction, and analysis. From here, I discuss the emergence of the Positivist-Inductive Approach, followed by the development of the Problem Approach. Then, I examine the movement towards Dependency/ World Systems theories. Finally, I highlight the contemporary issues raised in

Peace Research and conclude with an attempt to reflect upon and identify the theoretical directions most appropriate for this project.

THE PERIOD OF BORROWING

Since the pioneering days of Comparative Education, various attempts have been made to define the purpose and scope of Comparative Education. Historically, the beginnings of the field were not comparative at all (Hans, 1971). Studies were confined to description and collection of information on education in foreign countries. These works primarily dealt with what Kandel (1955) identified as the general anatomy of education: school organization, curriculum design, methods of instruction, time tables, and administrative practices. The main purpose of these initial studies was utilitarian; they were to provide a foundation for the reform of domestic education (Hans, 1971).

The first such study in the field is attributed to Marc-Antoine Jullien de Paris (Lauwerys, 1959; Bereday, 1964; Hans, 1971). In 1817, Jullien designed a comprehensive strategy to compare systems of education. In his paper, *L'esquisse et vues preliminaires d'un ouvrage sur l'éducation comparêe* [sic], Jullien conceived the purposes and methods of Comparative Education to include an "analytical" study of education in all countries with the goal of bettering national systems in accordance with local conditions (Hans, 1971).

Hans (1971) details similarly conceived reports that abounded in the 19th century. Matthew Arnold in England, Victor Cousins in France, Leo Tolstoy and K.D. Ushinsky in Russia, Domingo Sarmiento in Argentina, John Griscom, Horace Mann, and Henry Barnard in America all studied foreign educational systems with the explicit or implicit assumption that their respective country should borrow, albeit prudently, the best school practices from abroad (Bereday, 1964; Hans, 1971).

By focusing on the utilitarian applications, the scope of the pioneering studies was limited by a narrow concept of education. The principles underlying the development of national systems were barely considered (Hans, 1971). As such, these analyses, though not without interest, failed to contribute to a breadth of approaches to the problems of education (Kandel, 1955). Bereday (1964) identified the accomplishments of the 19th century as the first phase in Comparative Education and terms it the period of "Borrowing." During this period, emphasis was on cataloguing descriptive data and comparing the collected information in order to transport the best practices. Taking educational systems of one country and moving them wholesale to another was thought feasible in the 19th century.

THE PERIOD OF PREDICTION

The second phase of Comparative Education, which prevailed during the first half of the twentieth century, inserted a preparatory process before allowing any transplantation (Bereday, 1964). Bereday (1964) suggested that this era be called the period of "Prediction." The purpose of Comparative Education moved beyond borrowing to predicting the likely success of an educational system transplanted from one country to another.

The first attempt to predict and thus move away from a strictly utilitarian, borrowing approach was made by Sir Michael Sadler in his lecture, *How Far Can We Learn Anything of Practical Value from the Study of Foreign Systems of Education?* (Guildford, 1900, as cited by Kandel, 1955; Bereday, 1964; Hans, 1971; Mallison, 1975). Sadler urged students of Comparative Education to "try to find out what is the tangible, impalpable spiritual force which, in the case of any successful system of education, is in reality upholding the school system and accounting for its present efficiency" (Sadler, 1900, as cited in Kandel, 1955: 9). In this same lecture, Sadler conjured up the now famous garden metaphor to vividly illustrate that education is a living entity, not easily transplanted from one place to another (Kandel, 1955; Hans, 1971; Mallison, 1975). Sadler pushed scholars towards a more comprehensive point of view. He called for the acceptance of the principle that each educational system is not readily detachable but is instead fundamentally connected with the society that supports it (Bereday, 1964).

THE PERIOD OF ANALYSIS: THE HISTORICAL APPROACH

Many heeded Sadler's callings. The field of Comparative Education then moved toward a historical/cultural approach in which Friedrich Schneider and Franz Hilker of Germany, Isaac Kandel and Robert Ulich of America, Nicholas Hans, Vernon Mallinson, and Joseph Lauwerys of England were among the first to seriously consider the social foundations of education (Bereday, 1964). These scholars continued to redefine the purpose, scope, and method of Comparative Education but did so with a seemingly larger vision. The field of Comparative Education thus broadened to include a systematic investigation of other cultures and other models of education with the hope that extensive analysis would ultimately nurture greater sensitivity, lower the barriers of ethnocentrism, and contribute towards better international understanding. The work of Hans, Kandel, and Mallison exemplify the period.

Hans (1971) believed that national educational systems reflected both the future and the past. Factors that formed national pasts were seen as common to many nations. Therefore, the problems of education in different countries were somewhat similar. Moreover, he felt that national ideals of the future were the outcome of universal movements, and as such, the principles that guided their solutions could be identified and compared. For Hans (1971), the main purpose of Comparative Education became an analytical study of these national factors from a historical perspective and a comparison of the problems and solutions. He pronounced Comparative Education to be dynamic, forward looking, and intent on not only comparing existing systems but also on envisioning reform (Mallison, 1975).

Kandel was also reform minded. He compared various philosophies based not on theories but on actual prevailing practices. He paid special attention to nationalism and national character as a historical background to existing conditions (Hans, 1971). For Kandel, the study of foreign educational systems meant a critical challenge to one's own philosophy and therefore an ultimately clearer understanding of the basis underlying one's own educational system (Mallison, 1975).

Like Kandel, Mallison acknowledged that every definition of the purpose of education had an implicit philosophy. He, however, believed it was aimed at producing not the "natural man" but rather the kind of person that a particular society at a given period in history desires (Mallison, 1975). He focused on the identity and development of a "national character." Mallison (1975) concluded that the real purpose behind the study of Comparative Education was a systematic investigation of other cultures and other models of education in order to uncover the problems common to all. To identify the problems of education thus became a most important preliminary task (Mallison, 1975).

The historical/analytical period continued the prediction tradition, but it postulated that before prediction and consequent borrowing, systematization of the field must be in place in order to reveal the whole panorama of national practices in education. In order to aid this expanded vision, primary concerns began to revolve around analysis, theory, methods, and the clear formulation of comparative procedures (Bereday, 1964).

THE POSITIVIST-INDUCTIVE APPROACH

The positivist-inductive approach to Comparative Education developed from this stage. Pioneered by Bereday, Noah, Eckstein, and Husen among others (Hayhoe, 1986a), scholars were increasingly concerned with a theory of empirical investigation (Noah & Eckstein, 1969). During this period, scholars attempted to make Comparative Education "scientific" in order to support knowledge and beliefs and to eradicate the problems of bias and a lack of generally accepted criteria (Noah & Eckstein, 1969). Ideally, comparativists sought a method of inquiry that minimized the possibility of observer bias and maximized the validity of data. Above all, the method had to be self correcting and open to public scrutiny. All this pointed towards a method of science (Noah & Eckstein, 1969).

A scientific methodology offered a set of attractive tools for a rigorous quantitative analysis (Hayhoe, 1986a). Such a method for quantitative educational research could not only illuminate particular educational phenomena but also lead to a solidly empirical body of information about relationships between educational and societal factors across nations. On one level at least, comparisons could be made across nations within a common framework (Hayhoe, 1986a).

Bereday's work typifies the era. He advocated a "problem approach" (later taken up by Holmes) in order to develop a "total analysis." For him, analysis of the overall impact of education upon society in a global perspective was the culminating point of the discipline. He believed this final stage involved the formulations of "laws" or "typologies" that contributed to international understanding and a definition of the complex interrelation between the schools and society. The "total analysis" dealt with the immanent forces upon which all systems were built (Bereday, 1964). Bereday further asserted that as its final aim, Comparative Education should aspire to ease national pride in order to permit events and voices from abroad to count in the continued reappraisal and reexamination of schools.

THE PROBLEM APPROACH

Brian Holmes was largely influenced by the work of Bereday. Hayhoe (1986a) credits Holmes with developing the most thorough and credible critique of the positivist-inductive approach to Comparative Education. Holmes' methodologies differ, however, from the positivist approach in that the main notion of a predictive science of education does not seek the historical causes of educational phenomena but rather anticipates the likely consequences of various policy choices in specific societal contexts (Holmes, 1981, as cited by Hayhoe, 1986a).

Holmes (1965, 1981) worked on the assumption that "problems" arise out of asynchronous social change. Hence a theory of social change was needed if "problem" analysis was to be successful and replicable. He went on to assert that models and classificatory systems were also needed in the process of policy formulation, adoption, implementation, data identification, and outcome anticipation. To meet these needs, Holmes devised a system of classification. Based on Popper's theory of "critical dualism" or "critical conventionalism," Holmes proposed a hypothetico-deductive scientific methodology to view how scientific knowledge is furthered. Holmes, drawing upon Dewey, offered a highly specific definition of a problem. Holmes stated:

> In Dewey's conceptual framework, sociological laws are hypothetical policy solutions to identified problems. They are the basis on which the planned development of education should be built and, of course, they are statements which can and should be tested in experience. (Holmes, 1981: 80)

In addition to the quantifiable data placed within the "problem approach," Holmes proposed Weberian ideal types as a means of dealing with the cultural values that resist quantification yet are vital to an understanding of educational phenomena. The Problem Approach thus emphasized an understanding of the cultural values that create a context for educational phenomena.

Hayhoe (1986a), however, highlights its limitations. Implicit in both the Positivist Inductive and the Problem Approach philosophies is a belief in an objective reality. Deep-level cultural factors cannot be easily incorporated within this methodology. The validity of comparisons made between nations expressed in educational statistics is dubious. Hayhoe acknowledges that the Problem Approach does promote an understanding of cultural values; however, it also adheres to critical dualism. Hayhoe explains that an absolute distinction between facts and values prohibits scholarly contribution to normative choices, beyond the technical one of predicting results in specific societal contexts. The absolute adherence to a fact-value dichotomy and the lack of an integrative global framework limit the effectiveness of these approaches. Hayhoe underscores the need for an evaluative approach beyond predicting the outcomes of differing policy approaches. Moreover, she highlights the necessity for a framework which allows for the exploration of links between cultural educational phenomena and the global political economy (Hayhoe, 1986a).

THE DEPENDENCY/ WORLD SYSTEMS ANALYSIS

During the past twenty years, Dependency/World Systems Theories have emerged with certain characteristics to address these needs (Hayhoe, 1986a). Primarily based on Marxist dialectic thought and a historicist view, these theories propose that an accurate analysis of the historical process can provide normative directions. This lends the theory a certain integrity in contrast to the supposed value-free objectivity of the earlier approaches (Hayhoe, 1986a). The aim is toward a global systems theory in which educational phenomena in different countries can be understood in their relation to the international political economy (Altbach, 1980; Arnove, 1980, as cited in Hayhoe, 1986a). Moreover, this theory moves beyond the exclusive consideration of industrialized nations to include consideration of the Third World. The ties between industrialized nations and the Third World are discussed around the concepts of dependency and centre-periphery and are based on historical traditions, current economic realities, and the location of key educational and intellectual resources (Altbach, 1977). These socioeconomic theories provided new perspectives on the apparent failure of nonindustrial nations to benefit from educational transfers from developed countries (McLean, 1983). The dependency theory proposes that the nature of the structural relationship between the developed and less-developed states thwarts developments in education as well as in economic fields. These theories shift the emphasis from the "deficiencies" of Third World societies to a reevaluation of practices (McLean, 1983).

In an article which boldly calls for world-system analysis of education, Arnove (1980) defines its three central concerns: economic and cultural dependency, centre and periphery issues, and convergence and divergence in the international order. Arnove suggests that dependency theory delineates a descending chain of exploitation from the hegemony of metropolitan countries over peripheral countries, to the hegemony of the centres of Third World countries over their own peripheries. Arnove relates the concepts of centre and periphery to Wallerstein's notions of convergence and divergence in the global system. Arnove summarizes Meyer et al., to explain that convergence is produced as the world market and society subjugate all countries to the same force. The world market and society produce divergence by creating different roles for different societies in the world stratification system. For peripheral countries, partaking in the world system represents an opportunity for access to valued resources such as capital, technology, and skills; however, it also involves the risk of subjugation by stronger nations (Meyer et al., 1975: 233, as cited in Arnove, 1980: 49).

A world system analysis, however, does not confine these dynamics to an international capitalist order. Arnove (1980) stresses the need to recognize that knowledge and institutional exchanges occur not only between developed and less developed countries, but also among Third World countries. For example, India and the Phillippines host many international students from Southeast Asia, and India also provides scholarships to African students. Mexico and Argentina serve as a centre for Latin American scholars while students from the Middle East traditionally gravitated towards Egypt and until recently Lebanon (Altbach et al., 1985). Arnove (1980) points out that many countries change their position in the international stratification system. Countries can move from peripheral to centre status in the world system either by circumstances or choice. Certain Third World nations acquire prominence and become centres of hegemony themselves. A world system analysis allows for a certain flexibility in order to examine the flow of ideas and personnel within regional blocs of countries beyond the notions of First World, Third World. Arnove concludes with the conviction that historical studies of the effects of such shifts in a country's position in the world system would make a major contribution to an understanding of international education.

Other scholars are less convinced. In a thorough critique of dependency theory, McLean (1983) draws attention to the dangers of borrowing and applying economic theories to education. Economic reductionism and determinism obscure more than they reveal about the dynamics of cultural and educational dependency. Moreover, he states, these very theories have now been rejected by economists. He stresses that his criticism does not amount to a claim that the issues raised by dependency theorists are not important. On the contrary, McLean calls for a new analytical framework of greater applicability and flexibility in order to address these very same issues but with a keener sensitivity. The World Order Models Project (WOMP) offers such a framework with an avowed and explicitly normative perspective (Kim, 1979: 4).

WORLD ORDER MODELS PROJECT (WOMP)

WOMP expands the capacity of the dependency model beyond politics and economics to include a consideration of knowledge and culture. Kim (1979: 4) characterizes WOMP as a cross-cultural, transnational, and global research enterprise dedicated to the actualization of four central goals: one, the minimization of collective violence; two, the maximization of social and economic well-being; three, the realization of fundamental human rights and political justice; and four, the maintenance of ecological balance and harmony. Thus, motivated by the values of peace, economic well-being, social justice, and ecological balance, WOMP scholars raise culture to a place of significance in order to examine international academic relations in an interdisciplinary, global, and value explicit framework (Hayhoe, 1986a). Within this framework, WOMP theorists such as Johan Galtung (1975) and Ali A. Mazrui (1975a,b) acknowledge the strength of the periphery and its power to provide the means for a peaceful transformation of knowledge relations.

Johan Galtung

The work of Galtung (1975) complements the work of the dependency theorists and its critics. Concerned with the same issues, Galtung highlights the notion that premises are often as important as conclusions. His approach differs from a classical dependency/world system analysis by opening up the possibility of positive action in relation to global inequalities (Hayhoe, 1986 a,b). In an article on the possibility of a peaceful research methodology, Galtung (1975) conceptualizes structural violence based on four components of imperialism: exploitation, penetration, fragmentation, and marginalization. These components, he argues, apply to the relationship between the researcher and the researched. As summarized by Hayhoe (1986a: 70), the exploitation of the periphery by the centre in the vertical division of labour takes place with centre scholars creating theory and periphery scholars carrying out the more modest task of data collection or theory application. Penetration occurs to the degree that explanations or theories produced in the centre 'get under the skin' of periphery thinkers and researchers, establishing in the exploited a bridge of a 'local bourgeoisie' whose cultural alienation from their own periphery is thereby increased. Fragmentation happens to the degree that researchers in the periphery are separated from one another. They have close links to one or several centres but no channels of communication among themselves or with other peripheries. Thus, the possibility for an academic counter culture is obstructed. Marginalization results to the extent that the peripheral researchers remain in a permanent status as secondary scholars, dependent on and subordinate to the more influential scholars of the centre (Hayhoe, 1986a: 70).

To counter such imperialistic and oppressive practices, Galtung proposes a model of positive action, a model of nonviolent social science, based on four guiding principles: equity, autonomy, solidarity, and participation. Hayhoe (1986b) reflects on Galtung's alternative values:

> Equity suggests aims and forms of organization that are reached through full mutual agreement. Autonomy suggests a respect for the theoretical perspectives rooted in peripheral culture that would require center participants to gain a thorough knowledge of this culture. Solidarity suggests forms of organizations that encourage maximum interaction among peripheral participants and growing links between them and their fellow researchers. Participation intimates an approach to knowledge that does not stratify in a hierarchical way but assumes the possibility of a creative peripheral contribution from the very beginning. (Hayhoe, 1986b: 535)

Critics of Galtung argue that these formulations are 'naive' and expand the problem so broadly that it is impossible to develop guidelines and priorities for research (Falk, 1982: 147; Holsti, 1985: 60; Lawler, 1995: 74). Within this expanse, the implicit assumption may appear to be that the interests pursued by peace research are the expression of values held by specific groups of actors or investigators, dominant or otherwise (Lawler, 1995: 78). As Lawler (1995: 79) points out, this approach is in contrast to Galtung's work on structural functionalism where values are located within the subjective disposition of the members of a specific system to avoid the dangers of conservative functionalism whereby subjective values are characterized as objective needs of a system. These formulations overlook the problem of dominance and its bearing upon supposedly consensual social values (Lawler, 1995: 79). As such, the power, intellectual autonomy, and resistance of local subjects may be overshadowed. The work of Ali A. Mazrui (1975a, b), on the other hand, highlights the power of local subjects and offsets these potential risks.

Ali A. Mazrui

Mazrui (1975a), like Galtung, insists on positive action to global inequities. Mazrui strives to transform knowledge structures and consequently transform political and economic power structures. Mazrui likens the peripheral universities to a multinational corporation, thus linking cultural and economic dependency. For Mazrui, centre universities dominate periphery universities analogous to the ways in which multinationals dominate in economic matters. As such, the peripheral university epitomizes cultural dependency as the multinational epitomizes economic domination. Thus, both the African university and the African multinational face the same challenge: the process of decolonizing modernization without ending it (Mazrui, 1975a: 285).

Mazrui focuses on the potential of culture and knowledge in periphery nations to face this challenge. To this end, Mazrui proposes peripheral universities counter the domination of knowledge from the centre by the following three strategies: domestication, diversification, and counter penetration.

The first strategy seeks to domesticate imported knowledge and make it relevant to the local culture. The second strategy seeks to diversify the cultural content to create a truly inclusive global curriculum. The third strategy bids the African continent to counter penetrate, to reverse the flow of influence to western civilization (Mazrui, 1975a: 306-318). Mazrui warns that the first two strategies, the domestication of modernity and the diversification of its cultural content, cannot be fully realized without the third. Mazrui explains:

> Full reciprocal international penetration is a precondition for a genuinely symmetrical world culture. As Africa first permits its own societies to help balance the weight of western cultural influence, then permits other non -western external civilizations to reveal their secrets to African researchers and teachers, and then proceeds to transform its educational and intellectual world in a manner which makes genuine creativity possible, then Africa will be on its way toward that elusive but compelling imperative-not only to decolonize modernity, and not even merely to participate in it, but also to help recreate modernity anew for future generation. (Mazrui, 1975a: 317-318)

Mazrui's concluding remarks hold much promise for China's academic relations with African nations. In these programs, do African nations permit China to reveal their secrets? Have these programs begun to transform the international educational world? Have they moved Africa closer towards that compelling imperative of recreated modernity? These questions inform the larger objectives of this study as detailed in Chapter Four.

CONCLUSIONS

The values expounded by WOMP theorists, such as Johan Galtung and Ali A. Mazrui, provide the means to begin asking such questions and reflect upon the nature of international knowledge transfer. They further provide the means for placing such queries within a paradigm that is holistic, global, and transformative.

Knowledge paradigms do change. When Hans Weiler declared that the consensus over epistemology had eroded, he signified an emerging space for new ways of knowing. The growing importance of environmental paradigms, feminist paradigms, peace paradigms inevitably changes the linkages between knowledge and power relations. As the world shrinks, individual worlds expand. Human beings can now, perhaps more than ever, potentially see themselves in a world context. As a Canadian exploring the lives of African scholars in a Chinese context, this vision became critical. If the world has indeed become a "global village," then forms of knowledge, like nation states, must not be seen as isolated facts but as integrated wholes, adapted to accommodate all members (Masemann, 1990). Increasingly, the field of international academic relations is moving towards this understanding.

Notes

. For example, Altbach, Kelly, and Lulat (1985: 3) point out that while international centres of knowledge are now predominately based on a Western model, this was not always the case. As far back as 600 B.C., non-Western systems, such as the universities of Taxila and Nalanda in India, hosted foreign students and employed foreign faculty to teach in Sanskrit or Pali. Al-Azhar in Cairo still exemplifies the Islamic university, serving the entire civilization in the Arabic language. These international institutes were among the earliest known universities to which, at some points in history, Western scholars flocked.

CHAPTER FOUR

Research Design and Data Collection

INTRODUCTION

Within a theoretical paradigm informed by the work of Galtung and Mazrui of the World Order Model Projects, I proposed to examine the lives of African students in China. In January 1997, I responded to an announcement from the Education Office, The Embassy of the People's Republic of China in Canada (dated November 25, 1996) calling for research proposals for the China/Canada Scholars Exchange Program (Appendix B). In a letter dated March 25, 1977 (Appendix C), I was awarded the opportunity to conduct research in China contingent on final approval from Beijing. I was asked to resubmit my proposed plan of study in further detail. In a letter dated July 28, 1997 (Appendix D), I was asked to report to Tongji University in Shanghai by September 3, 1997. The letter and an Admission Notice (Appendix E) confirmed that the award had been granted. The award included tuition fees, on-campus accommodation, medical care, textbooks, living allowance, and international airfare. The award not only provided me with the means to do my research, but it also signified something more: official permission.

When the State Education Committee of the People's Republic of China officially granted me permission to go to China, I understood that I would carry out the approved research proposal. Upon arrival, however, key changes were made. The modifications altered the fundamental design and nature of the proposed investigation.

PURPOSE AND FOCUS

The purpose of this chapter is to trace the evolution of the changes in the research design and data collection procedures. To do this, I first outline my intended research protocol; second, I discuss the changes made to this protocol; third, I detail the actual protocol; fourth, I summarize the overall effects of these developments. Finally, I reflect on the methodological process and my personal location within this study.

PART ONE: INTENDED RESEARCH PROTOCOL

Intended Objectives of Study

In the proposals submitted to the Chinese Embassy in Canada and later to the State Education Commission in Beijing, I stated that the purpose of my study was to contribute to an understanding of international academic relations by highlighting a perspective rarely considered: the South-South dimension of the organization and transfer of knowledge. Specifically, I proposed to investigate the structures and patterns of Sino-African international educational exchanges. In addition I aimed to: .

i.   identify the nature and objectives of China's educational aid for Africa
ii.   identify African student priorities related to technology and training
iii.   investigate China's educational practices in response to those priorities
iv.   assess the extent to which programs were relevant and applicable to African conditions
v.   reflect upon the nature of South-South transfer of training and technology

I strove to contribute to an understanding of international development and training by directing attention towards the collaboration *between* Third World nations. With the help of many, I designed the overall research, instrumentation, and data collection procedures to meet these objectives.

Intended Methods of Data Collection

As stated, I intended to use quantifiable data, obtained from surveys of African students, as the framework for the more qualitative information, gathered through individual and group interviews with both African and Chinese parties.

Intended Time Line and Subject Populations

My intended time line was four months: September 1997 to December 1997. During this time, I planned to consult two subject populations in this study: African students and Chinese university staff. The African population included students ranging in age from approximately twenty to thirty years old and residing in China for at least one year in preuniversity language training. The Chinese population included faculty, administrators, and government authorities involved in international academic exchanges and foreign affairs.

Intended Sources of Data and Research Sites

I intended to collect data from three sources: student questionnaires, student interviews, and interviews with Chinese faculty, administrators, and government authorities. I intended to investigate three sites in two cities: Tongji University in Shanghai, Shanghai Medical University in Shanghai, and Zhejiang University of Agriculture in Hangzhou. These three sites were chosen to represent three different academic disciplines: Engineering, Medicine, and Agriculture, respectively.

Intended Procedures for Data Collection and Selection of Subjects

All African undergraduate and graduate students from the above three universities were to be asked to participate. For the survey, I proposed to go to the foreign student dormitory at each of the three sites. Once there, I planned to hand deliver a survey package to every African student in their choice of English or French. A Chinese version was also to be made available. I had hoped to contact at least one hundred students from each institute.

For the student interviews, participants were to have been selected from the same institutes. From among those students who volunteered an interest, I hoped to secure written consent for taped interviews from eighteen. I planned to select students according to their discipline, level, and geographical origin. I had hoped to interview six students from each of the three universities representing the three disciplines (Engineering, Medicine, and Agriculture). Within each discipline, I had hoped to speak with two students from each level of study (Bachelor's, Master's, and Doctoral). In all cases, I aimed to reach students from a variety of countries in order to achieve some geographical representation from the North, South, East, and West of the African continent.

For the staff interviews, participants were to be sought from these same three institutes. I also aimed to have six interviews from each school seeking input from various Chinese representatives: administrators in the foreign affairs office (*waiban*), officials at the student affairs' office (*liuban*), and professors. Whenever necessary, I planned to seek the assistance of an interpreter. Whenever possible, I hoped to use a tape recorder and secure written consent.

Administrative and Informed Consent

Administrative consent for the project had been secured from the Government of the People's Republic of China. Informed consent was to be secured from all individual participants. For the interviews, as mentioned, signed consent was to be sought. For the questionnaire, consent was considered implicit in its completion. The interview consent forms (Appendixes F and G)[1] and the first page of the questionnaire clearly emphasized the voluntary and confidential nature of the research.

Intended Instruments: Student Questionnaire and Interview Schedules

All research approaches endeavour to describe, explain, explore phenomena; as such, every method is only an approximation of knowledge, each providing a different glimpse of reality (Warick & Lininger, 1975; Babbie, 1973; Gray & Guppy, 1994). Each method has certain strengths and certain limitations as to the types of data, variables, and analytical approaches it generates.

As a guest in China, I felt the design of a questionnaire, as a prearranged program for collecting and analyzing information, would enable me to reach as many students as possible and would enable my hosts to know the questions I wanted to ask in advance. As Babbie (1973: 45) states, survey research is like a "crustacean: all the bones are on the outside." In other words, all the biases of survey research are more or less clear and upfront thus enabling others to appraise their implications. For these reasons, I chose to do a questionnaire (Appendixes H and I).

While the statistical approach of the questionnaire may provide descriptive material and suggest significant relationships between variables, it may lack the capacity to explore in-depth contextual factors. Such factors may be better addressed through the inclusion of qualitative interviews. Interviews offer a contrasting approach to the statistical analysis, providing a check on the findings and at the same time allowing "a conversation between the two approaches" to take place (Morgan, 1983, as cited in McMahon, 1988: 45). Thus, I combined both quantitative and qualitative methodologies in order to yield different types of data to provide a more comprehensive understanding of this study.

Questionnaire Design

The design of the questionnaire was the result of a truly collaborative effort. I am largely indebted to the academic works of Somsak Boonyawiroj (1982), Uko T.C. Ekaiko (1981), Yousef Feiz (1995), Cecilia Siqiong Huang (1994), Kang Ji (1993), Otto Klineberg and W. Frank Hull IV (1979), Georgine Konyu-Fogel (1994), Kathryn M. Mickle (1984), Shahrzad Poorshaghaghi (1992), John Porter (1962), and Hamdesa Tuso (1981). While all these studies dealt with some aspect of international student experience, and Ekaiko (1981) and Tuso (1981) dealt specifically with the experience of international students from Africa, all these works focused on international study in the North. For this reason, I relied heavily on the work of Otto Klineberg and W. Frank Hull's (1979), *At a foreign university: An international study of adaptation and coping*. In this study, researchers looked into the experience of foreign students from thirteen areas of origin, classified in the following divisions: Western Europe, United Kingdom, Eastern Europe, Black Africa, Arabic speaking countries, Iran, South Pacific countries, South Asia, South East Asia, Other Asian countries, Latin America, Canada, and the United States. These students were studying at universities in eleven countries: Brazil, Canada, France, Hong Kong, India, Iran, Japan, Kenya, United Kingdom, United States, and West Germany. The questionnaire in the study was not only applied internationally, it was constructed internationally. Throughout the construction process, the investigators "did not wish to adopt a Western, or an Asian, or any other ethnocentric approach" (p.10). Thus, they sought participation from scholars across the globe. While plans to include a country from the Communist world were not realized, they did receive input from nationals of Brazil, Canada, France, Ghana, Hong Kong, Iran, Japan, and the United States. The questionnaire in my project was largely influenced by this "international, interdisciplinary" study.

In addition to these academic texts, I also sought personal input from many people for a wide range of advice and perspectives. Yu Liming and Pei Chao assisted me in making the protocol and instruments appropriate to the Chinese context and sensibilities. Barbatus Gatoto, a Burundian who spent five years studying in China, was involved in the overall project from the start and contributed significantly to the construction of both the interview schedules and questionnaire design. For both instruments, he tailored the questions for an audience he knew better than I, and he translated the questions from English to French. I have named just a few individuals, but the methodology was the result of much effort by many.

Questionnaire Content

The questionnaire was designed to generate information about the lives of African students in China. The questionnaire package began with a letter to introduce myself, outline the nature of the project, and identify the purpose of the research. The letter requested students to participate in the study by completing the questionnaire and volunteering for in-depth interviews. Moreover, this letter stressed the voluntary and confidential nature of the research. Students were assured of anonymity, and those with any concerns were invited to contact me at my residence. The survey consisted of more than 125 both closed and open-ended questions. Most of the questions used a 5-point Likert scale and the remaining required a one word or short phrase answer. The questions were divided into eight sections: 1. Student Profile, 2. Motivation, 3. Issues, 4. Social Contact, 5. Academic Experience, 6. Chinese Language and Proficiency, 7. Financial Support, and 8. Future Plans. Section One was designed to generate a student profile by asking questions in the following areas: Biographical Information, Family Background, Academic Background, and Current Education in China. Section Two sought information about factors that motivated students to leave their country and make the commitment to study in China. Respondents were asked to rate the level of importance of thirteen factors that were focused around four areas of motivation: financial, academic, employment, and personal. The last question in the section was left open for students to comment on other factors that influenced their decisions. Section Three focused on issues that foreign students commonly face. Respondents were asked to rank their level of agreement on fourteen statements pertaining to student life. Again this section closed with an open-ended question to enable students to raise issues that I had not. Section Four dealt with social contact. Items in this section were designed to probe the sources of social companionship, close companionship, and frequency of contact with the host nation. Section Five dealt with academic experience. The first set of nine questions dealt with evaluating academic experience, and the second set of thirteen questions dealt with the relative ease or difficulty regarding academic tasks. Section Six dealt with Chinese language training, proficiency, progress, and the relationship among the three. Section Seven inquired about the sources, approximate percentages, and sufficiency of financial support. Section Eight asked students to reflect upon their future plans. The final section of the survey was open-ended, inviting students to provide any additional information. The first and the last page of the survey ended with a request for volunteers to contact me for an interview.

Intended Interview Schedule

For both sets of interviews, I had planned to use a semi-structured interview schedule centring around Galtung's formulations of equity, autonomy, solidarity, and participation as related to the three disciplines of Engineering, Medicine, and Agriculture. The intended interview schedule for the students (Appendix J) and the intended interview schedule for the Chinese participants (Appendix K) were designed to complement one another and reflect on the nature of the Sino-African exchanges in relation to the values and principles expounded by WOMP theorists. As discussed, this investigation was placed in a holistic, global, and transformative paradigm. As such, the four principles of Galtung's model of positive action (equity, autonomy, solidarity, and participation) and Mazrui's three strategies of African modernization (domestication, diversification, and counter penetration) were the theoretical directions in which I had hoped to shape the interviews and the entire study.

PART TWO: MODIFICATIONS TO INTENDED RESEARCH PROTOCOL

On September 3rd, 1997, I attended the first meeting with my assigned junior and senior supervisors. They were familiar with my intended protocol from copies of my two applications. I was asked to present an itinerary and submit my questionnaire. I did this immediately. On the 25th of September, I was called in for the second meeting. At this meeting, I was advised of the following two changes. First, I was asked not to distribute the questionnaire. I was permitted to administer the questionnaire but one at a time, and I was asked to remain present while students filled it out. Second, I was asked to do the questionnaire orally. I was asked not to have the students "check" their answers themselves; instead, I was to check their answers for them.[2] Beyond these changes in the method of questionnaire distribution and procedures, I later learned that the three planned research sites were no longer available.

Initial Impact of Changes

First, these changes alerted me to potential sensitivities surrounding the project. Though I had begun my research with a certain confidence (my proposals accepted, a scholarship granted, instruments in place), these new developments and their implications shook this assurance. Second, these changes signified that the choices regarding the time line, the research sites, and the methods for data collection were now to be arranged by my supervisors. I did not proceed until instructed to do so. Third, I went from a state of clarity (I knew what I had to do, how to do it, and how long I had) to a state of perplexity. I had understood that I had Governmental consent to investigate three sites, representative of three different academic disciplines. The focus of this investigation was to revolve around these institutes, these academic disciplines, and the African scholars and Chinese faculty found therein. Now my research would no longer revolve around three sites, three disciplines, and two subject populations. The change of the intended research sites consequently changed the intended subject populations, both of which ultimately changed the fundamental focus of my research. While I quickly understood that the premise of my investigation was to be transformed, I could neither predict nor control the degree or the direction of the transformation. While the intended design and methods for data collection were no longer acceptable, no alternative plans were advanced in their place. The sites changed, the focus disciplines changed, and the subject populations changed, yet I did not know beforehand what they had changed to. Thus, the core of my research problem and the framework in which I had envisioned looking at this problem had to be let go. In some ways, the project was now out of my hands. In the end, I did not systematically and consciously plan the itinerary, choose the sites, disciplines, population, and methods of data collection. In other words, I designed the original plan, but the actual plan was not fully designed by me. As the final protocol never fully revealed itself upfront, it is only now upon completion that I can reflect and reconstruct its unfolding.

PART THREE: ACTUAL RESEARCH PROTOCOL

Actual Time Line and Subject Populations

The intended four-month time line decreased to less than three. By the time I was authorized to begin data collection, five weeks had passed. My first assigned appointment with African scholars was more than one month after I had arrived: October 6, 1997. My first assigned appointment with Chinese parties was almost three months later: November 18, 1997. While the time decreased, the intended two subject populations (African students and Chinese university staff) increased to four (African students, graduates, Embassy Counsellors, and Chinese university staff).

Actual Sources of Data

The intended three sources of data (student surveys, student interviews, and interviews with Chinese staff) more than doubled. From the Chinese participants, I collected data from seven interviews with the following six participants:[3]

- Vice President, — University (two interviews)
- Director, Foreign Students' Office
- Deputy Dean/ Director, International Exchange Division, Associate Professor
- Professor, Department Head
- Director, Department of Foreign Affairs
- Acting Section Chief Engineer, Foreign Affairs Office.

In two cases, a translator was required, and each time an African student offered his skills. I also obtained one document from the State Education Commission, produced by the Department of Foreign Affairs, dated April 18, 1997 (Appendix L).[4]

From the African participants, I obtained one hundred and thirty-three questionnaires from students from twenty-nine countries and held seven interviews with the following three groups:

Students: one group interview with three (two undergraduates and one post graduate)
- two individual interviews with post graduates
Graduates: one group interview with four graduates from three Chinese universities. These graduates still lived in China but no longer studied
Embassy Counsellors: three individual interviews with Counsellors from three countries

In these seven interviews, I spoke with a total of twelve participants from six different countries. Only two of these participants were women. In addition to the questionnaire data and interviews, I obtained one document from the General Union of African Students in China (GUASC), Shanghai Executive Branch, dated July 4, 1996 (Appendix M) and one document from the foreign graduate students at Tongji University in Shanghai dated November 17, 1997 (Appendix N).

CHART 4.1
Overview of Actual Sources of Data

Sources of Data	African Participants	Chinese Participants	Comments
Questionnaires	133 questionnaires from students from 29 African nations, studying in over 12 faculties, across 14 sites, in 4 cities (Beijing, Hangzhou, Nanjing, Shanghai)		
Interviews	7 interviews (5 individual, 2 groups) with 12 participants from 6 different countries, across 5 sites in 3 cities	7 individual interviews with 6 people across 5 sites in 2 cities	14 interviews with 18 people (6 Chinese, 12 African) across 10 different sites in 3 cities
Documentation	Two pieces: Memorandum to the Resident State Education Commissioner, from the General Union of African Students in China (GUASC), Shanghai Executive Branch, dated July 4, 1996. Letter to the Director, Foreign Students' Office, Tongji University, from Master's and Doctoral Foreign Students, dated November 17, 1997.	One piece: Documentation produced by the Department of Foreign Affairs of the State Education Commission, dated April 18, 1997.	Three pieces in total.

Actual Research Sites

In total, the intended three research sites in two cities became eighteen sites in four cities. For the African populations, I visited the following fourteen sites in four cities:
- Ten universities in four cities for the student questionnaire
- One site from the above for the student interviews
- One site for the Graduate interview

- Three sites for the Embassy Counsellor interviews

For the Chinese populations, I visited the following six sites in two cities:
- Five universities in two cities for interviews (at three of these sites I did not meet with any students)
- The State Education Commission in Beijing for institutional documentation

CHART 4.2
Overview of Actual Sites of Data

	African Participants	Chinese Participants	Totals
Sites	14 sites in 4 cities	6 sites in 2 cities	18 different sites, in 4 cities: Beijing, Hangzhou, Nanjing, Shanghai

With the exception of the State Education Commission, the sites were chosen for me. Typically, my supervisors arranged an appointment and gave me a calling card of introduction. Upon my arrival at each institute, I presented this card to the Foreign Affairs Office.

Administrative and Informed Consent

Institutional consent was granted by way of these calling cards. Administrative consent from the Government of the People's Republic of China, though intact, did not imply individual institutional consent. Institutional consent was granted by some schools and not by others. In all cases, the nature of individual institutional consent varied greatly. Informed consent was secured from all individual participants. For the survey, consent was implicit in its completion. For the interviews, signed consent was sought, but oral consent had to suffice.

Actual Selection of Subjects and Procedures for Data Collection

There was no one approach for the selection of subjects and procedures for data collection. The Foreign Affairs Office at each institution proceeded in very different ways. Many African students were selected by the Foreign Affairs Office of their institute. Typically, the office arranged a room and chose students to meet me. In some cases, I was permitted to interview; in other cases, I was permitted to do the survey. In a few cases, I was permitted to do both. Some institutes asked me not to do the survey orally: "Why waste people's time?" Some institutes invited me back and allowed me to contact students independently.

For the interviews with African students, after a general invitation, all participants approached me. I waited for people to take up my invitation because I wanted them to feel confidence in me and comfortable with the project. Their level of confidence and comfort, rather than discipline, academic level, and geographical representation became the guiding criteria. In all cases, I obtained oral consent, and in all cases, I was permitted to tape the interviews. For the interviews with African Embassy Counsellors, I sought the assistance of a colleague who works in one of the embassies. He arranged all three interviews for me.

The Foreign Affairs Office of my host institution selected all Chinese participants. In all cases, I obtained oral consent, and in three cases, I was permitted to use a tape recorder. On the advice of my host institute, I approached the State Education Commission independently.

Impact of the Changes

While the time frame had already decreased from four months to three, the new procedures for data collection dramatically increased the required time. I had imagined I would distribute all of the surveys in one week and while waiting for the mailed returns begin the interviews. Now I had less time for interviews as I collected an average of ten surveys a week. Students' time commitments also increased. A thirty minute written survey easily became a ninety minute oral survey. The process not only required more time but also more formality. As students could no longer complete the survey alone, at their convenience, booking individual appointments was now necessary. Students committed a block of time, on a given day and hour.

Yet somehow, alongside the formality of these arranged meetings, alongside the formality of the oral protocol, an unexpected openness emerged. In many ways, the oral surveys resembled interviews and as such, provided an intimacy which may not have been realized in the intended format. I saw firsthand how students understood and responded to the questions. The oral surveys may have given students an opportunity to be more expressive and forthcoming, which ultimately provided more information and greater insight into their individual experiences. In the end, these oral surveys greatly influenced the actual interviews.

Actual Interview Schedules

As mentioned, the intended interview protocol was semi-structured to focus on academic issues for specific populations. As these populations changed and as the circumstances changed, this interview protocol changed in response. To accommodate any incoming, unknown participants, to provide maximum security and comfort for all, and to balance the formality and structure of the questionnaire process, I altered the semi-structured interview schedule to be as open as possible. I began every interview with both Chinese and African parties by asking one basic question: What are the essential elements in these exchanges?

I chose the expression "essential elements" to be as neutral as possible. I avoided words like concerns, issues, challenges as they might carry negative implications. Occasionally, participants asked for more of a lead. In these cases I asked: What are the essential elements of your experience? What are the essential elements in the lives of African students in China? What are the essential elements in these Sino-African educational programs? I asked them to imagine if they saw such a study in the library, what would they expect to find. What should be there? What should this investigation not miss?

In all cases, participants discussed only the issues that they chose to discuss. They told me what they wanted to and as much as they wanted to. I fully participated, but at the same time I deliberately and consciously said as little as possible to neither encourage nor discourage trains of thought. I attempted to put the interview agenda in the hands of the participants.

Reflections on the Interview Process

In the case of all interviews with African parties, the response was overwhelmingly positive. This may be attributed to the fact that I had a personal connection to the foreign student community through friends I had made during my previous stays in China. During those times, I was most closely linked to the Burundi community. When I initiated this project in 1995, I did so, as mentioned, with the assistance of one Burundian, Barbatus Gatoto, who now lives in Canada but had spent five years in China. As I was setting up the project, I sought further assistance by contacting my Burundi colleagues in China. Even before I made it to China, many students and one Embassy Counsellor already knew of the project. Once in China, my colleagues acted as my guides, accompanying me to all four cities. They introduced me and the project to the African community as a whole. Thus, these colleagues, many of whom had been in China for over nine years, garnered interest and, more importantly, trust from the larger African community. When doubt did arise as to whom I may be associated with, or where this data may end up, my colleagues were able to assure students, "Don't worry, I know her." With this assurance, participants needed no more leads. They knew what they wanted to tell me. Students, in some cases, actually sought me out to participate. When I arrived to do the interviews, they inevitably had prepared food for the occasion. One time, when my tape recorder broke down, I had to cancel an interview. When I did this, the student was anxious to reschedule: "Please, make sure I get my chance." When I thanked students for their participation, they in turn thanked me, not in response to my gratitude but for "granting" them an interview. Participants wanted to talk. Interviews ran from two to four hours as students strove to articulate their full experience. They also strove to ensure I understood. On all 'essential elements', the message was clear. When I did not immediately understand some accompanying points, students took pains to explain the details to me. For example, when one group spoke about the image of South Africa, I was perplexed. As a Canadian, I associated South Africa primarily with apartheid. My bias was admittedly negative. However, they were talking about a view of South Africa that associated this nation with a high level of development. The bias towards South Africa was positive. I had never regarded South Africa in this light before. When these types of gaps arose, participants worked hard to help me to understand. In some cases, they wanted to compare their perspective, their experience with mine. Above all, African participants had a sincere desire to be heard and understood.

In comparison, Chinese participants were more reserved. On three occasions, interviews were cancelled at the last minute, and on two occasions, the person just did not show up. When interviews were held, they ran from forty-five to ninety minutes. In all but one case, participants looked to me throughout the interview to provide leads. In these cases, I offered a single word, a short sentence to move the conversation along. This reservation may be attributed to several factors. First, I may have had a professional connection to the Chinese participants, but I did not have a personal one. Unlike my connection with the African community, I had no "insider", so to speak, to assure participants of confidentiality. Moreover, all the interviews were set up by my host institution. Thus, with an official record of our contact, none of the interviews felt strictly private. This was very clear in one case where the interview was held with three other people in the room, one of whom was taking photographs. Finally, it is worth repeating that I conducted these interviews in English (and in two cases with an African student as an interpreter). While all but these two participants were completely fluent, English was, nevertheless, not their first language. If these interviews had not been limited by my lack of fluency in the Chinese language, participants may have been more at ease and the flow of communication may have been greater. These reasons may begin to explain why these participants were comparatively more reserved and reluctant to move beyond the surface of an official discourse.

As this study is participant driven, the findings reflect the level of participation of each group. As a result, the findings are largely the voices of the Africans. In the interviews, African participants spoke volumes, and on the questionnaire, they wrote copious notes. Many students even attached separate pieces of paper to the already lengthy document.

In all cases, I hope that both Chinese and African readers will find a full and accurate portrayal of their testimony. And while each group painted a sharply contrasting picture of their relationship, these pictures did not necessarily conflict because they did not fully overlap. In the interviews with the Chinese parties, most felt the "essential elements" related to the history of the program. In addition to the history, the most commonly mentioned elements were language and scholarship aid. In the interviews and on the questionnaires, the African parties felt that the "essential elements" related to race, social contact, and funding. These essential elements became the essential elements of this investigation. Their focus became my focus.

CHART 4.3
Overview of Actual Interviews with African Participants

PARTY	NOTES
Students	4 interviews (2 group, 2 individual) with 9 students from 4 countries. All interviews taped.
1. Group of Three	Group interview with two undergraduates and one post graduate
2. Post Graduate A	Individual interview with one post graduate
3. Post Graduate B	Individual interview with one post graduate
4. The Four Graduates	Group Interview of 4 students who graduated from 3 different Chinese universities, still living in China but no longer studying.
Embassy Counsellors	3 individual interviews with 3 counsellors from 3 countries. All interviews taped.
5.Embassy Counsellor A	
6. Embassy Counsellor B	
7. Embassy Counsellor C	
COMMENTS	In total, 7 interviews with 12 people from 6 different countries were held at 5 sites in 3 cities from October to December 1997. The exact dates and sites have been omitted to ensure confidentiality.

CHART 4.4
Overview of Actual Interviews With Chinese Participants

PARTY	NOTES
1. Vice President, —-University	Recorded (individual interview with photographer & 3 office assistants)
2. Director, Foreign Students' Office	Recorded
3. Deputy Dean/ Director International Exchange Division, Associate Professor	Written
4. Professor, Department Head	Recorded
5. Director, Department of Foreign Affairs	Written
6. Acting Section Chief Engineer, Foreign Affairs Office	Written
7. Deputy Dean/ Director International Exchange Division, Associate Professor (*same person, second interview)	Written
Comments	In total, 7 individual interviews with 6 people were held at 5 sites in 2 cities from November to December 1997. The exact dates and sites have been omitted to ensure confidentiality.

PART FOUR: SUMMARY

To summarize, I believe the following points bear emphasis. First, the changes made to the questionnaire distribution and procedures ultimately required more time and reduced the total number of anticipated participants. I had left Canada with six hundred and fifty surveys and returned with one hundred and thirty-three completed. I had hoped to interview thirty-six people, but in the end, I interviewed eighteen (twelve African parties and six Chinese parties). Second, the changes made to the sites attenuated the focus of the project; the attenuation of the focus shifted the intended concentration of the investigation. The loss in numbers and focus, however, brought forth unexpected gains.

As the intended protocol indicates, I planned to go to three sites, in two cities, and concentrate on three disciplines. As the final protocol reveals, I actually visited eighteen sites, in four cities, and looked at over twelve disciplines. As the number of sites, cities, and disciplines increased so then did the variety of subjects. Within these three institutions, I intended to consult with two subject populations: African students and Chinese authorities. In the end, I met with three groups of African participants inside and outside academic institutions: students, graduates, and Embassy Counsellors. Moreover, I met with Chinese administrators, faculty, and government officials, not from three but rather from six institutes. In the case of both subject populations, the numbers of participants may have decreased, but the variety and scope of the populations broadened. The change of sites ultimately resulted in a greater number and variety of sites, cities, disciplines, and populations. Thus, paradoxically, the conditions placed upon the intended protocol brought forth unexpected expansions and directions to the project.

CHART 4.5
Overview of Intended and Actual Research Protocol

	INTENDED PROTOCOL	ACTUAL PROTOCOL	COMMENTS
TIME	Four months: 09-12 1997	Three months: 10-12 1997	The intended 2 subject
SUBJECT POPULATIONS	1. African Students	1. African Students	populations increased
		2. African Graduates	to 4.
	2. Chinese University Staff 1. African Students	3. African Embassy Counselors	
		4. Chinese University Staff	
		The intended 2 subject populations increased to 4.	
SOURCES OF DATA (See Chart 4.1 for more information)	Interviews • 18 African Students • 18 Chinese Staff Student Surveys: 300(min)	Interviews • 5 African Students • 7 Chinese Staff Student Surveys: 133 total NEW ADDITIONS Interviews • one group interview with 4 graduates • 3 individual interviews with 3 Embassy Counsellors Documentation: • State Education • Commission • African Student Union • Graduate Students at Tongji (Appendixes L- N)	The intended three sources of data more than doubled
SITES (See Chart 4.2 for more information)	For Both Populations: 1.Tongji University 2. Shanghai Medical University 3. Zhejiang University of Agriculture	For African Populations: Interviews: 5 different sites in 3 cities Surveys: • 10 different sites in 4 cities For Chinese populations: • 6 sites in 2 cities (5 universities & State Education Commission)	The intended 3 sites in 2 cities became a total of 18 different sites in 4 cities: Beijing, Hangzhou, Nanjing Shanghai
INSTRUMENTS	1. Student Survey 2. Structured Interviews (Appendixes H-K)	1. Student Survey 2. Open Interviews	Interviewees asked to reflect on "essential elements".
CONSENT	• Informed consent from all institutional and individual participants. • Implicit consent for completed surveys • Signed consent for interviews (Appendixes F, G)	Informed consent from all institutional and individual participants. • Implicit consent for completed surveys • Oral consent for interviews	Administrative consent from the Government, though intact, did not imply individual institutional consent. The nature of individual institutional consent varied.

PART FIVE: FINAL REFLECTIONS ON METHODOLOGY

To close this chapter on methodology, I would like to reflect upon my personal "set of coordinates," as outlined in Chapter One. Discourses of identity are, of course, not universal. Thus, it is difficult to speculate and offer broad generalizations as to how I, as the researcher, may have been perceived by participants from thirty different countries (twenty-nine African nations and China). What I can discuss more confidently is my own location within the study.

As mentioned, this study is informed by a recognition of the reflexive character of social research; that is, the acknowledgement that both the researcher and the researched are engaged in the construction of knowledge (Borman, Goetz, & Le Compte, 1986; Davies, 1982; Eisner, 1991; Hammersley & Atkinson, 1983; Joron, 1992). In other words, the researcher, as the instrument that engages the situation and makes sense of it, is not detached from the process but rather an integral part of it (Eisner, 1991). Many argue that this understanding is not just a matter of methodological commitment but rather existential fact.

> There is no way in which we can escape the social world in order to study it; nor, fortunately, is that necessary. We cannot avoid relying on 'common-sense' knowledge nor, often, can we avoid having an effect on the social phenomena we study. (Hammersley & Atkinson, 1983:15)

Thus, rather than engaging in attempts to eliminate the effects of the researcher, this principle grounds personal experiences as a starting point (Hammersley & Atkinson, 1983). An experientially grounded approach locates the researcher in the research process by examining what was initially a personal reaction to a given situation (Joron, 1992).

My Location

My personal reaction to this situation under study was indeed a strong impetus. For over two years, I travelled throughout Asia and lived in China. This life contrasted profoundly with anything I had ever known. Immersed in this ancient state, I began to understand the relative degrees of modernity which distinguish Canada. While the contrast between Chinese and Canadian living was becoming clearer to me, I was also learning much about other cultures. For a long portion of my stay, I lived with foreign students in a university dormitory. These students were predominantly African. We were all foreign to each other, different cultures, religions, languages, races, and an infinite range of social practices all coexisting and interacting. I began to see how others live and learn. This study is, in part, my attempt to capture some of what I saw.

My personal reaction to this situation was one of compelled interest and admiration. China has been offering educational opportunities to citizens of many African nations for over forty years. During this period, young African men and women have made profound commitments and sacrifices in pursuit of higher education. In doing so, they have achieved significant accomplishments. My initial interest and admiration only intensified over time and was later coupled with a sense of considerable gratitude. I feel truly grateful to have had the opportunity to go to China in the first place. Once in China, I feel I had the further privilege to encounter an extraordinary group of scholars in a remarkable situation. My initial impetus for this study was to communicate what I could about this situation, to acknowledge the sustained cooperation between China and Africa, and to pay tribute to all those involved.

Subjectivity

Because every individual's history and hence world is unique, what moves one person, what one sees, how one reacts and interprets a situation inevitably bears an "individual signature" (Eisner, 1991). Hammersley and Atkinson (1983) draw attention to the need to recognize and make a personal mark:

> Once we abandon the idea that the social character of research can be standardized out or avoided by becoming a 'fly on the wall' or a 'full participant', the role of the researcher as active participant in the research process becomes clear. He or she is the research instrument par excellence. The fact that behaviour and attitudes are often not stable across contexts and that the researcher may play an important part in shaping the context becomes central to the analysis. (Hammersley & Atkinson, 1983: 18)

Subjectivity is thus not viewed as a liability or contamination; rather, subjectivity is regarded as a matter of standpoint, individual input, and challenge to the idea of pure objectivity which assumes that the object can and ought to be separated from the subjects (Cook & Fonow, 1990, as cited in Joron, 1992). Scholars across the disciplines have come to reject the myth of a context-free and value-free location from which an objective way of knowing can be established and instead recognize that every way of knowing, every form of knowledge is organized from somewhere, a null point, a set of coordinates that mark the standpoint of the knower (Schutz, 1962, as cited in Jackson, 1991).

My Standpoint as Knower

As an occidental woman from the Western world attempting to conduct research, primarily in English, about the experiences of a diverse African scholarly community in the People's Republic of China, my "way of knowing" was both necessarily informed and limited by my "set of coordinates." Though much remains undefined in these formulations, variations in such coordinates inevitably affect the type of questions asked, the choices made, and the interpretations given in the research process (Borman, Goetz, & Le Compte, 1986; Pennycook, 1992; Woolgar, 1988).

One goal of the reflective research process is to attempt to extract, to try to understand the situation from the perspectives of others. This goal raises inherent challenges as it conflicts with the assumption held by many field researchers that we can best represent but never truly know the condition of others. Elliot Liebow reflects on this viewpoint:

> This perspective — indeed, participant observation itself — raises the age old problem of whether anyone can understand another or put oneself in another's place. Many thoughtful people believe that a sane person cannot know what it is to be crazy, a white man cannot understand what it is to be black, a Jew cannot see through the eyes of a Christian, a man through the eyes of a woman, and so forth in both directions. In an important sense, of course, and to a degree, this is certainly true; in another sense and to a degree, it is surely false, because the logical extension of such a view is that no one can know another, that only John Jones can know John Jones, in which case social life would be impossible. I do not mean that a man with a home and family can see and feel the world as homeless women see and feel it. I do mean, however, that it is reasonable and useful to try to do so. Trying to put oneself in the place of the other lies at the heart of the social contract and of social life itself. (Elliot Liebow 1994: xiv-xv, as cited by Bailey, 1996: 113)

Despite the inevitable challenges, looking beyond ourselves, beyond the limits of our own experience is, of course, a necessary and valuable exercise. In this process, to "actively refuse" an "authoritative voice", I have attempted to identify my possible set of coordinates (hooks, 1988: 42-48). Meaning in this study is constructed and shaped, in part, by the tools I have chosen and know how to use. This meaning then reflects not only qualities "out there" but also the tools and appreciation I bring along (Eisner, 1991).[5] Thus, I conclude this chapter on methodology and begin the presentation of the findings with the acknowledgement that my location necessarily informs, limits, and implicates the knowledge production of this study.

Notes

.¹ The design of the interview consent forms was largely adapted from Ashbury, F. D. (1991), *International scholarly exchange and status recognition: a case study of China's exchange scholars and students* (pp.225-226). Unpublished doctoral thesis, York University, Toronto, Ontario.

.² These two changes were further modified by my host institution and the institutions I visited. In the end, the conditions changed over time and from place to place.

.³ Job titles were obtained from the participants' business cards.

.⁴ I met with a representative of the State Education Commission on December 15, 1997 to seek documentation and an interview. An interview was not granted, but on December 19, 1997 this document was left for me to pick up. It was subsequently translated by Zhang Xiaoman in Shanghai, January 1998 and Pei Chao in Montreal, June 1998. The original version and translated summary appear in Appendix L.

.⁵ This being said, readers might begin to wonder whether I have excluded comments or perspectives that did not fit the analysis because, at times, this analysis may seem one-sided and overwhelmingly negative. However, no positive or alternative points of view were omitted. I made a full and conscious effort to incorporate *all* written questionnaire comments and as much of the interview material as I possibly could. This task was made easier by the fact that students wrote and spoke in a remarkably consistent manner and voice. Their message, although at times possibly discomforting, was strong, clear, and unambiguous.

Chapter Five

Student Profile

INTRODUCTION

In earlier chapters, the historical, theoretical, and methodological foundations for this investigation have been considered. Before turning to the findings, it bears repeating that the scholarship programs for African students
to pursue higher education in China were established as early as 1949, as
China, inspired by the spirit of Bandung and a revolutionary zeal, made a
concentrated effort to establish close ties with the newly independent
African nations. Part of this effort included establishing educational ties, a
bond which has been sustained ever since. A document produced by
China's Ministry of Foreign Affairs of the State Education Commission
April 18, 1997 (Appendix L), itemizes Sino-African exchanges and divides
the cooperation into four time periods: 1949 to 1966, 1966 to 1978, 1979
to 1989, and 1990 to 1996. During these four periods, China has sent more
than ninety delegations, dispatched more than 400 teachers, and provided
various kinds of educational support to over twenty-five African countries.
In addition, China has hosted more than eighty delegations from African
nations and admitted approximately 4,570 African exchange students
(Appendix L).[1] This study looks at the experience of 133 of these students
in China today.

ORGANIZATIONS OF FINDINGS

The findings of this study are organized into seven chapters according to
the seven sections of the questionnaire: Student Profile, Motivation, Issues,
Social Contact, Academic Experience, Chinese Language and Progress, and
Financial Support.[2] Within each chapter, I follow the general format of each
questionnaire section. In all cases, I use the quantifiable data, obtained
from the questionnaire, as a framework in which to place the more quali-

tative information, obtained through individual and group interviews with African and Chinese parties.

Section One of the questionnaire was designed to generate a profile of the students in this study by asking questions in the following five areas: Biographical Information, Family Background, Cultural Background, Academic Background, and Current Education in China. Therefore this chapter, the first of the seven chapters of findings, is divided into the above five parts.

PART ONE: BIOGRAPHICAL INFORMATION

The biographical section sought information in six categories: nationality, sex, age, religion, marital, and parental status of students. A total of 133 students filled out the survey, 53 (39.8%) in English and 80 (60.2%) in French. Of these 133 students, 6 declined to identify their nationality. From the remaining 127 respondents, twenty-nine nations were identified, as presented in Table 5.1.

TABLE 5.1
Nationality of African Students Who Completed Questionnaire by Frequency and Percentage (%)

Nation	Frequency	%
Burundi[3]	24	18.0
Cameroon	9	6.8
Congo	8	6.0
Rwanda	8	6.0
Mali	7	5.3
Uganda	6	4.5
Equatorial Guinea	6	4.5
Ghana	6	4.5
Guinea Conakry	6	4.5
Sierra Leone	5	3.8
Kenya	4	3.0
Namibia	4	3.0
Tanzania	3	2.3
Gabon	3	2.3
Madagascar	3	2.3
Mauritius	3	2.3
Zambia	2	1.5
Benin	2	1.5
Ethiopia	2	1.5
Zaire	1	0.8
Botswana	1	0.8
Chad	1	0.8

Continued on next page

TABLE 5.1
Nationality of African Students Who Completed Questionnaire
by Frequency and Percentage (%)

Nation	Frequency	%
Lesotho	1	0.8
Mozambique	1	0.8
Niger	1	0.8
Nigeria	1	0.8
Somalia	1	0.8
Sudan	1	0.8
Togo	1	0.8
No Response	6	4.5
Total	133	100

Gender Distribution

Of the 133 respondents, the vast majority, 85.7 % (114), were male while 14.3 % (19) were female.[4] Of the twenty-nine countries represented, women came from twelve. Men and women from Tanzania and Mauritius were equally represented (two out of four and one out of two respectively). The single student in the sample from Botswana was female, as was the single student from Lesotho. From Cameroon, four out of nine students were female; from Namibia, three out of the seven students were women; from Rwanda, two out of seven; from Kenya, one out of five; from Congo, one out of eight; from Sierra Leone, one out of five; from Guinea Equatorial, one out of six; and from Burundi, one out of twenty-four students was a woman.

Age

The age distribution of all survey respondents ranged from 21 to 43; the mean was 29.1 years old; the median 29.0. The age range of women was considerably younger (21 to 33); the mean and median age of the women was 25.0.

Religion and Marital Status

Of the 126 people who answered the question on religion,[5] the majority of respondents, 75.2 % (100), identified themselves as Christian; 19.5 % (26) identified themselves as Muslim. The majority of students, 74.4% (99), were single: 82 male and 17 female. The total percentage of married respondents was 24.8 % (33), which broke down into 31 males and 2 females. Of those 24.8% (33) who were married, 21.2% (7) had spouses in China while 63.6% (21) did not (the remaining 15.15% (5) did not answer the question). The majority of respondents, 63.2% (84), did not have children while 24.8% (33) indicated that they did. Of those with children, 75.75% (25) did not have their children with them in China while 12.12% (4) did (the remaining 12.12% (4) did not answer the question).

PART TWO: FAMILY BACKGROUND

The Family Background section of the questionnaire sought information about the education and occupation of parents, number of siblings, birth position,[6] and location of growing up.

Parents Formal Education

On an ascending scale, Table 5.2 indicates the highest level of formal education of respondents' parents.

TABLE 5. 2
Highest Level of Formal Education of Father and Mother by Frequency and Percentage

FATHER	Frequency	Percentage	MOTHER	Frequency	Percentage
Bachelor's	26	19.5	Secondary	34	25.6
Elementary	24	18.0	No Formal	27	20.3
Post Secondary	23	17.3	Elementary	25	18.8
Secondary	18	13.5	Post Secondary	25	18.8
No Formal	15	11.3	Bachelor's	11	8.3
Master's	14	10.5	No Response	9	6.8
No Response	8	6.0	Master's	2	1.5
Doctoral	4	3.0			
Other	1	0.8			
Totals	133	100		133	100

Most students indicated that the level of their Father's formal education was higher than their Mother's. While 11.3% (15) of Fathers had no formal education, almost double that figure, 20.3% (27), of Mothers had no formal education. Moreover, this data indicates that while there was considerable range, more than half the students came from family backgrounds where at least one parent had obtained some formal higher education. Over half of students' Fathers, 50.3% (67), and almost one quarter of students' Mother's, 28.6% (38), had obtained some form of higher education. Compare these figures with the regional data for Sub-Sahara Africa provided by *UNESCO Statistical Yearbook* 1989 and 1995. This data reveals that in 1960,[7] 0.5% males and 0.1% females were enrolled in Higher Education, (0.3% of the total population). These UNESCO figures reveal, by contrast, that the students in this study came from families that had obtained very high levels of formal education.

TABLE 5.3
Enrollment Ratios for Higher Education (Third Level)[8] in Sub-Saharan Africa[9] 1960-1993

Year	Total	Male	Female
1960	0.3	0.5	0.1
1970	0.8	1.2	0.3
1975	1.1	1.7	0.4
1980	1.6	2.5	0.6
1985	2.2	3.4	0.9
1986	2.2	3.4	0.9
1987	2.2	3.5	0.9
1990	3.0	4.1	1.9
1992	3.3	4.5	2.1
1993	3.4	4.7	2.2

Sources: UNESCO Statistical Yearbook, 1989, Table 2.10, pp.2-31, 2-23 and 1995, Table 2.10, pp.2-26, 2-27.

Parental Occupation

For parental occupation, I assigned value labels and categorized respondents answers into ten fields, as displayed in Table 5.4.[10] Table 5.5 presents the percentage of respondents' answers that fell into each of these ten categories.

TABLE 5.4
Assigned Categories of Student Response for Parental Occupation

Assigned Categorical Label	Student Responses
Business	Banker
	Businessman/ Entrepreneur
	Economist
	Factory Owner
	Personnel and Administrative Manager
	Self Employed
	Senior/ Certified Accountant
	Trader
Education	Education Inspector
	Education Officer
	Professor
	Proprietor of Islamic Schools
	Teacher
Farming	Farmer
	Peasant
Government	Agricultural Officer
	Assistant of Rural Development
	Civil Servant
	Customs Officer
	Court Chairman
	Diplomat
	Mayor
	Military Officer
	Minister of Foreign Affairs
	Police Officer
	Politician
	State Agent
	Traffic Manager
Housewife	Housewife
Professional	Doctor
	Engineer
	Nurse
Secretarial	Secretary
Skilled Labour	Auto worker
	Carpenter
	Chauffeur
	Fitter
	Mechanic
	Phone Exchange Technician
	Plumber
	Printer

Continued on next page

TABLE 5.4
Assigned Categories of Student Response for Parental Occupation

Assigned Categorical Label	Student Responses
Skilled Labour	Traditional Chef
	Tailor/Dressmaker
Other	Imam
No Answer	Blank
	Deceased
	Retired

For Father's occupation, respondents' answers fell into eight of these categories: 20.3% (27) Government, 14.3% (19) Farming, 11.3% (15) Education, 10.5% (14) Business, 9.0% (12) Skilled Labour, 6.0% (8) Professional, 0.8% (1) other, and 27.3% gave no answer or wrote "retired" or "deceased." For Mother's occupation, respondents' answers fell into nine categories: 30.1% (40) Housewife, 10.5% (14) Farming, 9.8% (13) Education, 7.5% (10) Business, 6.8% (9) Government, 5.3% (7) Secretarial, 3.8% (5) Professional, 3.8% (5) Skilled Labour, and 22.6% (30) gave no answer, or wrote "retired" or "deceased." Table 5.5 highlights this data.

TABLE 5.5
Occupation of Father and Mother by Frequency and Percentage

FATHER	Frequency	Percentage	MOTHER	Frequency	Percentage
No Answer	37	27.8	Housewife	40	30.1
Government	27	20.3	No Answer	30	22.6
Farming	19	14.3	Farming	14	10.5
Education	15	11.3	Education	13	9.8
Business	14	10.5	Business	10	7.5
Skilled Labour	12	9.0	Government	9	6.8
Professional	8	6.0	Secretarial	7	5.3
Other	1	0.8	Professional	5	3.8
			Skilled Labour	5	3.8
Totals	133	100		133	100

The two top categories for Father's occupation were Government 20.3% (27) and Farming 14.3% (19) while the two top categories for Mother's occupation were Housewife 30.1% (40) and Farming 10.5% (14). While 'Farming' may signify an occupation held by wealthy, land owning proprietors, the students' frequent use of the word "peasant" to describe their parents' occupation suggested to me that this was not the case. This information seemed to indicate that African students in China came from diverse socioeconomic backgrounds.

PART THREE: CULTURAL BACKGROUND

This section sought information about students' level of exposure to different cultures while growing up and through previous travel. This information may be indicative of students' ability to adapt to a new culture. The majority of respondents, 55.6% (74), indicated they had previously travelled outside their country while 43.6% (58) had not. Of those that had gone abroad, most, 27.8% (37), had travelled to other African countries while 14.3% (19) had travelled to both African and European countries, 8.3% (11) had travelled to Europe, 3.8% (5) had travelled to Asian countries, 0.8% (1) had travelled to North American countries, and 0.8% (1) had travelled to North American and European countries. For 28.6% (38) of respondents, the stay was more than six months, but for 25.6% (34) the stay was less than six months.

On a scale from 1 to 5, respondents were asked to rank their level of personal exposure to six other cultures while growing up. The largest percentage of responses to the question, *while growing up, how much exposure did you have to other cultures?* , fell into the 'none' category for each culture listed: 49.6% indicated that while growing up they had 'no' exposure to South American/Caribbean cultures, 44.4% had 'no' exposure to *Asian* cultures, 39.8%   had 'no' exposure to Middle Eastern cultures, 34.6% had 'no' exposure to North American cultures. On the other hand, 32.3% had 'great' exposure to other African cultures while 36.8% had 'moderate' exposure to European cultures. This data is presented in Table 5.6.

Finally,   to get a sense of students' overall cultural exposure, all the counts from the given variables of each culture (African, Asian, Middle Eastern, European, North American, South American/Caribbean, and Other) were collapsed into one variable of 'Overall Exposure', as presented in Table 5.7. As the cumulative totals reveal, while growing up, only 4.1% had 'very great' exposure to other cultures while the majority 54.7% (509 total count) had 'no' exposure to other cultures while growing up.

TABLE 5.6
Level of Personal Exposure to Other Cultures while Growing Up by Percentage (%) and Frequency (FQ)

	Very Great		Great		Moderate		Small		None		No Response		Totals	
	%	FQ	%	FQ	%	FQ	%	FQ	%	FQ	%	FQ	%	FQ
Other African	15.0	20	32.3	43	27.8	37	9.8	13	8.3	11	6.8	9	100	133
Asian	1.5	2	3.8	5	8.3	11	24.1	32	44.4	59	18.0	24	100	133
Middle Eastern	0.8	1	5.3	7	9.8	13	21.8	29	39.8	53	22.6	30	100	133
European	9.0	12	15.0	20	36.8	49	18.0	24	11.3	15	9.8	13	100	133
N. American	1.5	2	6.8	9	16.5	22	23.3	31	34.6	46	17.3	23	100	133
S. American/ Caribbean	0.8	1	1.5	2	9.8	13	16.5	22	49.6	66	21.8	29	100	133

TABLE 5.7
Overall Exposure to Other Cultures (Collapsed Variable) by Cumulative Percentage (%) and Counts (C)

	Very Great		Great		Moderate		Small		None		Totals	
	%	C	%	C	%	C	%	C	%	C	%	C
Overall Exposure	4.1	38	9.2	86	15.8	147	16.2	151	54.7	509	100	931

PART FOUR: ACADEMIC BACKGROUND

The Academic Background section of the questionnaire sought information about students' secondary schooling and higher education prior to their studies in China. These questions also sought to determine the extent to which students felt informed about their academic and living conditions in China prior to their departure.

Secondary Education Prior to China

When asked about the type of secondary school attended, the majority, 63.9% (85), indicated they had gone to public school; 22.6% (30) attended missionary schools; 7.5% (10) attended foreign schools; and 6.0% (8) attended private schools. Of these schools, the majority, 50.4% (67), were boarding schools; 38.3% (51) were day schools; and 6.8% (9) were both. In 70.7% (94) of the cases, respondents were not the first in their family to go to secondary school while 28.6% (38) were the first. The majority, 57.9% (77), were not the first in their family to receive higher education while 42.1% (56) were the first.

Higher Education Prior to China

The majority of students, 64.7% (86), attended university before going to China; 33.8% (45) had not. Of those that attended university before going to China, 38.3% (51) held a Bachelor's degree; 11.3% (15) held a Diploma; 6.0% (8) held a Master's degree; and 3.0% (4) had incomplete studies (some university education without completing a credential). Of those that had previously attended university, the majority had attended in their own country. A small minority of students, 7.5% (11), had attended university in another country. As Table 5.8 reveals, most of these students, 72.7% (8), who had attended university in another country did so on the African continent. This table offers just a glimpse of international academic exchanges within Africa. Beyond the continent, one student from Zambia had gone to Britain, one student from Equatorial Guinea had gone to Spain, and one student from Guinea Conakry had gone to Russia.

TABLE 5.8
Country of University Attended Prior to China

Country of Origin	Country of University
Burundi	Zaire
Congo	Sierra Leone
Congo	Zaire
Equatorial Guinea	Spain
Guinea Conakry	Russia
Mali	Senegal
Mali	Senegal
Nigeria	Morocco
Rwanda	Congo
Zaire	Congo

Level of Prior Information

The last question in this section of the questionnaire inquired about the extent to which students felt informed about the academic program and living conditions in China prior to their departure. Using a five-point scale from 'very well informed' to 'not informed at all', only 12.8% (17) felt 'very well' or 'adequately' informed while 23.3% (31) chose the middle option, indicating they felt 'fairly well informed'. The overwhelming majority, 62.4% (83), felt 'not adequately informed' or 'not informed at all'.

Students were asked to name the source of their information. Answers were classified into the following four sources: the Chinese Government, their own Government, other students, and the media. Most respondents, 38.3% (51), wrote that they received information from the Chinese Government. Of the 38.3% (51) who received information from the Chinese Government, the majority 66% (34) indicated they were 'not adequately' or 'not informed at all', though one student, 0.8%, who specified that he was 'very well informed' was informed by a 'Chinese teacher' in his country. Of those informed by their own Government, 17.3% (23), the majority, 65.2% (15), felt they were 'not adequately' or 'not informed at all'. Of those informed by other students, 19.5% (26), the majority 73% (19) felt they were 'not adequately' or 'not informed at all'. Of the 12.0% (16) informed by the *media,* the majority, 37% (6), felt 'not adequately' or 'not informed at all.' Regardless of the source, the majority of students indicated they were 'not adequately' or 'not informed at all'. A Zambian student summed up these sentiments:

> Students coming to study in China are not given adequate information, and they have to find out many things by themselves. This situation puts many students at a loss when they arrive because they do not know what they have to do. (Survey Note)

In fact, this was just one of many students' comments about the inadequacy of prior information. These comments will be discussed later in the findings.

PART FIVE: CURRENT EDUCATION IN CHINA

The final part of the student profile sought information about students' current studies in China. Students were asked to identify their faculty, specialization, level of study, number of program years, their year, and the total number of years they had been in China. As Table 5.9 and 5.10 indicate, the 133 respondents in this study came from twelve different Faculties and sixteen Majors.

TABLE 5.9
Current Faculty of Study in China by Frequency and Percentage

Faculty	Frequency	Percentage
Agriculture	37	27.8
Engineering	34	25.6
Political Science	15	11.3
Architecture	11	8.3
Science	7	5.3
Pharmacy	6	4.5
Economics	6	4.5
Management	5	3.8
Education	3	2.3
Law	3	2.3
Arts	2	1.5
Mathematics	1	0.8
No Response	3	2.3
Totals	133	100

TABLE 5.10
Current Major/Specialization by Frequency and Percentage

Major	Frequency	Percentage
General	25	18.0
Chemistry	21	15.8
International Relations	14	10.5
Civil Engineering	12	9.0
Environmental Studies	9	6.8
Food Science and Technology	9	6.8
International Commerce	7	5.3
Computer Science	6	4.5
Textiles	6	4.5
Urban Studies	5	3.8

Continued on next page

TABLE 5.10
Current Major/Specialization by Frequency and Percentage

Industry	5	3.8
Radio Engineering	3	2.3
History	2	1.5
Transportation	2	1.5
Rural Economics	1	0.8
Sociology	1	0.8
No Response	5	3.8
Totals	133	100

On average, a Bachelor's degree in these fields takes five years of study: one year of language training and four years of the subject. Most Master's and Doctoral programs are three years in length, and of the three people studying for a diploma, two indicated their program was 1.5 years in length, and one indicated his program was just five months long.

As Table 5.11 indicates, the majority of students in this study, 64.6% (86), were at the graduate level: 2.3% (3) were working towards a post graduate diploma; 53.4% (71) were working towards a Master's degree; and 9% (12) were working toward a Doctoral degree. The remaining 33.1% (44) of students were undergraduates.

TABLE 5.11
Current Degree Sought in China by Frequency and Percentage

Degree Sought	Frequency	Percentage
Post Graduate Diploma	3	2.3
Bachelor's	44	33.1
Master's	71	53.3
Doctoral	12	9.0
No Response	3	2.3
Totals	133	100

Most students, 42. 1% (56), who answered this questionnaire were in their first year; 30.8% (41) were in second year; 18.0% (24) were in third year; 1.5% (2) were in fourth year; 0.08% (1) were in fifth year; and 5.3% (7) had graduated from their particular program of study. Overall, respondents of this survey had lived in China from 5 months to 13 years, the mean being 3.29 years and the median 2.5. years.

PART SIX: SUMMARY PROFILE

A total of 133 students from twenty-nine African nations filled out the questionnaire. The majority of respondents, 85.7% (114), were male and 14.3 % (19) were female. The majority, 74.4% (99), were single and 24.8% (33) were married. The age of respondents ranged from 21 to 43 years, the average was 29, and in general, most had limited exposure to other cultures while growing up. Students came from diverse socioeconomic backgrounds, but the majority had at least one parent who had obtained a formal higher education. The majority of students, 64.7% (86), first attended university in their own country, and of these most, 38.3% (51), held a Bachelor's degree. Prior to their departure, the majority, 62.4% (83), felt they were not adequately informed about their academic program and living conditions in China. The majority, 53.3% (71), were currently studying for their Master's degree, and the largest concentration of students was in the fields of Agriculture 27.8% (37) and Engineering 25.6% (34). On average, students who answered this questionnaire had lived in China for 3.29 years. The next chapter discusses the factors that motivated these students to go to China in the first place.

Notes

.[1] Unfortunately, I was unable to find out exactly how many African students were in China at the time of this study. Figures and trends from secondary sources differ. Ten years ago, Cheung's (1989: 32) observed that in China, "The proportion of African and Third World students are falling. From making up the vast majority of foreign students in the 1970s, the 1,500-odd African students today account for one-quarter of the total." While Cheung reported an overall decrease in enrollment, Sautman (1994: 416) reported a time of increase, "In 1982, more than 400 of the 1,800 foreign students in China were Africans. By 1988, there were some 1,500 Africans among 6,000 foreign students." At this time, Delf (1989: 12) and Cheung (1989: 16) also estimated that 1,500 Africans were studying in China. And two years before this, Scott (1986b: 20) reported, "Some 1,600 of the 3,500 foreign university students in the country are from African countries." Scott's 1986 figure of 3,500 foreign students differs substantially from Sautman's 1988 figure of 6,000. Sautman (1994:425) further estimated that Africans are less than one-tenth of one percent of [all] university students in China."

.[2] The last section of the questionnaire, Future Plans, did not garner a significant response. The information that it did produce proved to be beyond the limitations of this study, and thus I chose not to include it as part of the findings.

.³ Burundi has a comparatively high representation because, as mentioned, Burundians assisted me in all stages of this project. These colleagues garnered interest and participation from the larger African community and from their own community in particular.

.⁴ Regional data for Sub-Sahara Africa provided by *UNESCO Statistical Yearbook 1995*, Table 2.10, pp.2-26, 2-27, provides an interesting comparison for these figures. This UNESCO data reveals that in Sub-Sahara Africa 1993, 3.4% of the total population was enrolled in higher education, with 4.7% of the age cohort male and 2.2% of the age cohort female. Thus, women's representation in higher education was approximately half of men's,

(46.8%), or, in other words, for every two males approximately one female was enrolled. In this study, women's participation in higher education in China was 14.3%, a considerably lower representation than their participation in higher education across the Sub-Sahara.

.⁵ From here on in, the numbers reported are indicative of the number of students who answered the particular question. Thus, in the cases where not all 133 students replied to the question, the total counts do not add up to 133, but the percentages were still calculated from 133.

.⁶ Number of siblings and birth position in the family did not prove to be significant factors for students.

.⁷ Given that the average age of students in this study is 29.1, it may be reasonable to assume that many of these parents were of the age to be enrolled in institutes of Higher Education around 1960.

.⁸ UNESCO defines Third Level education to include universities and other institutions of higher education. They specify that the data in these tables are standardized and based on the current national system of education of each country (UNESCO Statistical Yearbook, 1995, Table 2.10, pp.2-26, 2-27).

.⁹ This data excludes Arab states of the Middle East.

.¹⁰ In retrospect, I would have asked this question differently. First, I should have asked the question in the present and in the past tense. Many students wrote that their parents were "deceased" or "retired". These answers, along with blanks, I had to classify as 'No Answer'. Second, I should have asked students their parents' occupation and then asked the students to categorize the occupations themselves according to a list of categories. I did not do this and so was left with the task of trying to classify occupations across many diverse cultures. After deliberations and consultations, I finally sorted student responses into the following categories as presented in Table 5.4, recognizing they may not be entirely fitting.

CHAPTER SIX

Motivation

Section Two of the questionnaire sought information about factors that motivated students to study in China. On a five-point scale from 'Very Important (1) to Very Unimportant (5),' respondents were asked to rate the level of importance of each motivating factor. Thirteen potential factors were grouped around four areas of motivation: Financial, Academic, Employment, and Personal.[1] The last question in the section was left open for students to comment on other factors that influenced their decision to study in China.

PART ONE: FINANCIAL MOTIVATIONS

Three factors centered around potential financial motivations. *Financial assistance from family* was 'not applicable' for 21.2% (28) of students, 'unimportant' for 12.0% (16), and 'very unimportant' for 46.6% (62). The *level of financial assistance in China compared to home* was 'not applicable' for 20.3% (27), 'unimportant' for 20.3% (27), and 'very unimportant' for 25.6% (34). *Obtaining a scholarship,* however, was not only the greatest motivator in the financial category, but it was also the most often cited factor among all four categories of motivation. An overwhelming majority of students, 82.8% (110), rated this factor as 'very important' and 'important'. Table 6.1 highlights this data.

TABLE 6.1

Financial Motivations

Frequency (FQ) and Percentage (%) of Respondents who indicated the
Level of Importance of Each Factor

	Very Important		Important		Not Applicable		Unimportant		Very Unimportant		No Response		Totals	
	FQ	%	FQ	%	FQ	%	FQ	%	FQ	%	FQ	%	FQ	%
1. I obtained a scholarship.	55	41.4	55	41.4	10	7.5	3	2.3	9	6.8	1	0.8	133	100
2. I received more financial aid by studying in China than at home.	6	4.5	16	12	27	20.3	27	20	34	25.6	23	17.3	133	100
3. My family promised me assistance if I studied abroad.	2	1.5	4	3	28	21.1	16	12	62	46.6	21	15.8	133	100

Obtaining a Scholarship: The Enabling Factor

An abundance of written comments supported and supplemented these statistical findings. Two students specifically pointed out the financial opportunity that the scholarship provided. For one Burundian,[2] the "scholarship
provided financial independence" while the scholarship enabled a student
from Congo "to avoid financial difficulties." Other students commented
on the academic opportunity that the scholarship provided. A Kenyan student was "very keen to pursue a post graduate degree, . . . and a Chinese
scholarship opened the door first." A Namibian woman "just wanted education no matter where in the world, . . . [and] China [came] up first." A
Zambian student further explained that "scholarships to study abroad are
really hard to obtain, and when I was offered a Chinese scholarship I had
to take it up" (Notes from five surveys).

The difficulty of getting a scholarship and the sense of obligation to
"take it up" was echoed repeatedly. At the same time, the majority of students specifically made a point of highlighting the fact that China was not
their first choice but their only choice to study abroad. A student from
Madagascar wrote, "I came to China because I had a scholarship in China,
and I didn't find any scholarship in other foreign countries." A Mauritius
woman repeated this sentiment, "I wanted to go study abroad, but China
would have been the last country I would have chosen if I had the means
to pay for my studies myself. The bursary was definitely the decisive factor." A student from Benin noted, "I wanted to study in any other country,
as long as it was more developed than my own country, and I didn't get any
scholarship other than the Chinese one." For a Guinean, the Chinese scholarship was also "the only factor that made me come." Finally, a man from
Cameroon indicated that China was his only choice and last resort for
higher education. He explained that, "after having tried several times to get
a scholarship to the developed countries such as France, Belgium,

Germany, and Canada without any success, I tried China and a scholarship was given to me, and since I absolutely wanted to study abroad, I came to China" (Notes from five surveys).

The strongest motivator was the pursuit of higher education; students "absolutely wanted to study abroad." Students chose to pursue higher education abroad rather than higher education in China, per say. Many, like this student from Sierra Leone, were "looking for an opportunity to study abroad no matter which country." Though China may not have been the first choice, for many it was their only choice. A student from Congo acknowledged, "It's only China that gives graduate scholarships," and for one Burundian, China also provided the "only possibility" to pursue his "dreams":

> Since a long time ago, I dreamed of studying agriculture and coming here was the most certain way of achieving that. In fact, in my country, I wasn't sure of being oriented to the faculty of agricultural studies. Also, I wanted to study abroad, and China was the only possibility available. (Survey Note)

China may have been the "only possibility available" because in some cases this "only" choice was made available through a meritocratic process. A man from Togo pointed out that the Chinese scholarship was his "only chance" because "they give a test, and it's not dependent on relations, [and] since I only had my head, and I passed the test, so here I am." A student from Mali also earned the scholarship with his "head." He wrote, "In the last year of secondary school, the first three students in grades after the first semester had the right to a scholarship abroad, . . . [when I got one] . . . I didn't hesitate to go to China." A Burundian added, "When I finished the university, I was the first in my class, so I was able to go study in China" (Notes from six surveys). In these cases, students emphasized that the scholarship was an earned opportunity.

A comment from a student, in China since 1985, seemed to sum up a number of these issues:

> You know that in Third World countries, young people have major problems when they want to discover the world. So by coming to China with the help of the Chinese scholarship, it allowed me to do my studies and at the same time see the world. However, if I had got another opportunity, I would have gone to any other country. (Survey Note)

Thus, out of all thirteen factors from all four categories, this first factor, *obtaining a scholarship*, was by far the most often cited motivator. Obtaining a scholarship seemed to be the most important motivating factor in the sense that it was the enabling factor, making all other factors secondary.

The Chinese Scholarship

Two of the three African Embassy Counsellors that I spoke with reported that their countries receive ten annual scholarships from China. Though as little as three years ago, the number of scholarships used to be twice as many, these Counsellors remarked that their country receives more educational scholarships from China than any other country. One Counsellor added, "I don't think we ever sent ten students to America or France." Moreover, these Counsellors noted that the scholarships from China were not only "more in number . . . [but also] more regular . . . [and] more reliable." As such, their Governments considered the Chinese scholarships to be "very valuable," and they "count[ed] on" them in their annual planning.

Process of Recruitment and Selection

One Counsellor detailed the process of recruitment and selection:

> When I get the ten scholarships here [in Beijing] in February, I send the forms directly to the Ministry of Education [in my country]. They advertise openly . . . in the local papers, on the radio. [They announce], we have ten scholarships from China [so] go to the Ministry and choose the course. (Embassy Counsellor A)

The same Counsellor reported that "out of the announcement for ten you get over one hundred applicants, . . . [so] the process of selection is quite competitive." Interested candidates must go to the Chinese embassy and apply. While these scholarships used to target undergraduates, they are now primarily for graduate students. Post graduates and those "coming to do the arts . . . or humanities, just apply direct"; however, for undergraduates this application involves an exam. The Counsellor explained that for undergraduates interested in "science related courses, engineering, medicine, actuarial science, agriculture, . . . [the Chinese embassy administers a written exam], equivalent of A-levels in maths, physics, and chemistry." The Counsellor added that he thinks this exam is "unnecessarily difficult [and thus] eliminates many . . . [if not] most." After the exam, the Chinese embassy "short lists" those with the highest marks and then "according to the facilities [in China], they select their ten. They give them their tickets, and [students] come" (Embassy Counsellor A).

Despite initial interest, however, the ten scholarships are often not filled. The Counsellor holds his Government and the students responsible:

> It's on the fault of the inefficiency of our Government and, of course, our own students. . . . Some pull out at the last minute, and we can't replace them. . . . In fact, for the last three years, that is how it has happened. You have planned for ten, and then at the last minute, when they are going on the plane, you find somebody bringing their tickets back saying, 'I am not going'. (Embassy Counsellor A)

Finally, on behalf of those that do go, the Counsellor expressed his gratitude for the individual opportunities and national links the scholarship
provided:

> As far as my country is concerned, we are quite happy that we have ten
> scholarships every year from the Chinese government. So we feel grate-
> ful. That is somebody to educate ten people for you at university level
> and now postgraduate [level]. That's great, that's quite great because
> our university intake back home, for instance, is how many? Not very
> many. . . . So you get ten extra through. . . .
>
> If you calculate how much you would spend on a student per year. . .
> . And that student is probably going to study six years, if they are com-
> petent enough for post graduate [study]. . . . And that one won't affect
> another ten coming in. . . . So we actually have a cumulative number,
> so it's quite good.
>
> We are very grateful for that cooperation, . . . [and] that's good enough
> to maintain a link between the two countries. The scholarships, . . . on
> that level, definitely . . . enhance our relations. It's one of the levels of
> cooperation actually. (Embassy Counsellor A)

Preference for the North
While grateful for the enhanced bilateral relations and the individual
opportunities, the Counsellor did confirm and elaborate on the strong preference, among African students, parents, and governments, for scholarships to the North. The Counsellor stated:

> You will always find, there is a preference. . . . Given the chance, the
> students will always choose to go to the North. . . . Coming to China,
> . . . or going to Russia, or Cuba, we also sometimes get those scholar-
> ships, you find that they are not well liked. . . .These South-South rela-
> tions. . . . It's not the same as if you are going to study in the Sorbonne
> or in the University of Toronto. So there is that kind of attitude.
> Definitely there is a preference. . . . That, I am sure of. (Embassy
> Counsellor A)

From the point of view of the students, the Counsellor identified four main
reasons "students are attracted" to the North:

> [One], any scholarship from the North is always good, quite good
> actually, very good, generous really. If it's a good scholarship, the
> allowance will feed you properly, dress you properly, and at the end of
> the day, maybe he will go home with a car if he saved properly. So the
> North scholarships are generous really, . . . that's for sure. That's quite
> sure.

Two, the language is so clear to them. . . . They are teaching in a language that is quite familiar to them. . . . You are not going to the hassle of going to learn Russian first, or Spanish, or Portuguese here, . . . or Chinese here. . . . Relatives and friends, the first thing they ask is, 'Oh, you are going to China, how are you going to learn Chinese . . . ? How are you going to learn Chinese fast?' . . . They don't have to go into the hassles of learning Chinese in Canada or in America or in France. They know French already, or they know English already, so they go straight into the system they know.

Three, in our education systems back home, we in Africa, . . . have already [leaned] towards those who have colonized [us]. So if it is a British system, I mean, if they are colonized by Britain, they are already following a British system, colonized by France, following a French system back home, Belgium, the Belgium system. Germany was deprived of their colonies, so nobody is following that system. And the America system is almost suited to all. So . . . it's just a continuation really, from nursery school to university, just entering a classroom and you are meeting the same lecturer who is teaching you what you expect, straight, right, direct from high school. So they settle down very quickly. [Four], the facilities, the libraries, the books, the student interaction, [are all] excellent also, so they definitely choose that. They always choose that [Northern] one. (Embassy Counsellor A)

Thus, while "students are attracted" to the North because of the generosity of scholarships, familiarity of the language, familiarity of the educational system, and the level of facilities, for the students' parents, the preference for an education in the North is related to social status:

From the point of view of the parents, you know, we in the Third World, especially the Africans . . . who have conservative, old parents . . . have always dreamed to go where the colonial masters are. Just go there and come and boast that you have been there. Kind of a higher social status.(Embassy Counsellor A)

From the point of view of the home Government, the Counsellor identified two main reasons for a preference towards an education in the North: speed and quality.

From the point of view of the Government, they would prefer the North. One, because [students] go there, and they finish faster and come back to work. . . . Two, there is even that slight inclination that they are even getting [a] better education because they have better facilities. . . . Definitely, we would be happier if we got more scholarships from the North. (Embassy Counsellor A)

Despite this strong preference among African students, parents, and Governments, scholarships to the South nevertheless do enhance bilateral national relations and provide more opportunities for individual students to pursue studies abroad. As the Counsellor stated, North or South, "you are having a chance to go to university anyway." For many students, having the chance to go to university meant the chance to fulfill scholarly goals, as revealed in the next section on Academic Motivations.

PART TWO: ACADEMIC MOTIVATIONS

In this section, students were asked to consider three potential factors in the area of Academic Motivations. As Table 6.2 reveals, students had little *fear of not being accepted into a university at home,* as the majority, 59.4% (79), cited this as an 'unimportant' and 'very unimportant' motivating factor. However, a total of 47.4% did cite *China's better facilities* as a 'very important' and 'important' motivating factor.

TABLE 6.2
Academic Motivations
Frequency (FQ) and Percentage (%) of Respondents who indicated the Level of Importance of Each Factor

	Very Important		Important		Not Applicable		Unimportant		Very Unimportant		No Response		Totals	
	FQ	%	FQ	%	FQ	%	FQ	%	FQ	%	FQ	%	FQ	%
4. In my country, I feared I would not be accepted into a university.	3	2.3	3	2.3	28	21.1	9	6.8	70	52.6	20	15	133	100
5. In China, facilities in my field of study were better than in my country.	30	22.6	33	24.8	16	12	16	12	25	18.8	13	9.8	133	100
6. In China, there were courses and facilities not available in my country.	37	27.8	36	27.1	16	12	9	6.8	26	19.5	9	6.8	133	100

Facilities in China

While some students "just wanted to get another degree" or "just wanted to further [their] education," most students went to China with specific academic goals that required specific facilities. For example, a man from Mali wrote, "After my bachelors in Biology, my dream was to continue my studies in relation to the fixation of N2 or in another domain in Sol Microbiology, but in my country, continuing my studies was difficult . . . because of the reasons in 5 and 6 [facilities]. So I was a volunteer in my university lab, and I found the Chinese bursary, and I came." A woman from Rwanda also indicated that she came to China for "advanced" facilities. She wrote, "After the international conference on women which was carried out here, I think that China was an appropriate place for me to do my specialization, and . . . in the fields of education, China is somewhat better advanced." While these students commented about *China's better facilities*, the majority of students, 54.9% (73), also indicated that *facilities and courses unavailable in their own countries*, was a 'very important' 27.8% (37) and 'important' 27.1% (36) motivating factor. A woman from Cameroon noted that her "main purpose" in coming to China was academic. She wrote, "I was interested in the environmental sciences, particularly in waste and waste water management and treatment technology. There doesn't exist an environmental department in my university of origin; consequently, I was very happy when I had the chance of coming to China." Likewise, a Rwandan student of Chemical Engineering wrote, "I wanted to study in China because [that] faculty . . . does not exist in my country. And the Chinese technology is not bad in the world." For other students, their chosen faculties were in place at the undergraduate but not at the graduate level. One student wrote, "The lack of an opportunity to do my post university studies in my country made me accept to go study anywhere there was such a possibility." Similarly, a student from Rwanda noted that he was "in China for my graduate studies . . . because in my country, that level doesn't exist until now" (Notes from six surveys).

China's Academic Reputation

Beyond these three factors listed in the survey, students raised many other issues pertaining to academic motivation. Several students were motivated by the reputation of specific Chinese institutes, academic specialities, and scholarly traditions. A Zambian man wrote that "Tongji University is renowned for its programs in Architecture, and given the rate at which the property construction industry is growing and the role this university is playing, I thought it would be better to undertake this programme here to obtain vast practical experience which I can apply back home or elsewhere." A student from Guinea Conakry looked to Chinese scholars and hoped to "benefit from the experiences of the eminent Chinese researchers in order to make my country benefit from them." A Nigerian man found "The best place to study traditional Chinese pharmacy is China" while a Kenyan student looked to learn from China's accomplishments. He noted, "China is the most populous nation in the world and has a rich cultural history. It is able to feed its vast population. It is therefore an opportunity for an agriculturalist to appreciate its intensive agricultural production system"(Notes from six surveys). While these students had clearly defined academic goals, others were motivated to go to China for employment related reasons.

PART THREE: EMPLOYMENT MOTIVATIONS

As Table 6.3. reveals, the largest response to all three factors in the category of Employment Motivations fell into the 'not applicable', 'unimportant', and 'very unimportant' rankings.

TABLE 6.3
Employment Motivations
Frequency (FQ) and Percentage (%) of Respondents who indicated the
Level of Importance of Each Factor

	Very Important		Important		Not Applicable		Unimportant		Very Unimportant		No Response		Totals	
	FQ	%	FQ	%	FQ	%	FQ	%	FQ	%	FQ	%	FQ	%
7. In my country, a degree from China is worth more than a local degree.	6	4.5	9	6.8	37	27.8	13	9.8	52	39.1	16	12	133	100
8. I felt I could get a good job at home with a degree from China.	9	6.8	19	14.3	33	24.8	17	12.8	46	34.6	9	6.8	133	100
9. I felt I could get a good job abroad with a degree from China.	4	3	19	14.3	41	30.8	15	11.3	39	29.3	15	11.3	133	100

Students overwhelming disagreed that *a degree from China was worth more than a local degree*: 27.8% (37) cited this as 'not applicable', 9.8% (13) 'unimportant', and 39.1% (52) found this 'very unimportant'. In the margin beside this factor, one student wrote in capital letters "NOT TRUE," punctuated with three exclamation marks. This notation summed up the majority opinion on this factor.

Employment Prospects

A few students commented about the relationship between studying in China and employment prospects. One Doctoral candidate from Uganda studying Construction Management "wanted educational qualifications to get a position back in my government." A Kenyan studying for his Master's in Agriculture felt that "Improving academic qualifications would be a step in my upward mobility as far as (job) opportunities are concerned." For a Tanzanian woman, "China was the best way to obtain more education for my future career as an architect. I wanted to be more prepared to face various challenges ahead of me." Also looking ahead was an Architecture student from Gabon for whom "the most important [motivating] factor [was] related to . . . the possibilities which [will] open up . . . after training" (Notes from four surveys).

These four comments were atypical. No other student wrote about employment motivations. For most students, obtaining a higher degree was important for employment prospects, but obtaining a Chinese degree was not directly related to future opportunities. The majority of students, 72.2% (96), indicated that *getting a good job at home* or *getting a good job abroad*, 71.4% (95), was neither an applicable nor important motivation for going to China. One Zambian student clarified this point. Marked by an asterisk in the margin, he wrote that it was "not really a degree from China but a higher degree [that] will get me a good job." He emphasized his point on the bottom of the page stating that "The most important factor was the need to obtain a higher degree which will really make a difference in job opportunities" (Survey Note).

Ghana: A Unique Case

In all of this, one exceptional case emerged. Ghanaian students revealed that their motivation for going to China was almost exclusively related to employment. Ghanaians neither "apply" nor "even initiate" going to China but instead are nominated by their employer for job related training.

Nominated Government Workers

These students explained that China's ten annual scholarships go straight to the Scholarship Secretariat Body. This secretariat then contacts government institutions, such as research establishments, universities, and science and technology corporations, such as water and sewage plants, electrical plants, and highway boards. Supervisors in these institutions then nominate candidates to go to China.

Seniority and Timing

Nominations are based on seniority, and the country of your nomination is a matter of chance. One student explained that in his company, every individual has an opportunity for training. According to a seniority list, workers are automatically nominated when the offer comes from the secretariat. He stated, "It just happened that I was next on the list to be trained abroad, and the opportunity was to go to China." As another student put it, "When your time comes, [they] call you. It's not that they say, you go to China, you go to the US, you go to Japan, . . . but the scholarships, as they come, they give." In other words, "It's a matter of . . . what comes your way at what time." If these workers had been in a different place at a different time, a scholarship may have been granted to a different country. One student explained the process, "When you are nominated, your name is sent to that [scholarship] secretariat . . . from there . . . [you are] given the application forms, and you fill them in." However, students felt that this application was more like an invitation, and this invitation was really more like an obligation.

Obligations

As one student put it, "When your boss calls you and tells you that they are nominating you, who the hell are you to say that I won't go? . . . In one way or the other, you feel obliged because you have to get your Master's at the place we work, . . . the minimum is a Master's." A second student also shared this sense of obligation, "Actually, you don't have any option, you don't have a choice." He continued, "I am due for a promotion, and before I move on to the next level I should have the Master's degree. . . . I was next on the list, so I had no option than to take the offer." These students did concede that technically "Nobody will force you to come," but by refusing the first offer, it may be years before a second offer is made, if it is ever made at all. In either case, workers risk losing promotions.

In addition to these risk factors and the sense of obligation, age was another issue. One student felt that, "Age is very important . . . [because] most of the scholarships . . . specify age limits. . . . If people are younger, in their twenties and teens, they have more chances." Another felt that "Age was running out for me to wait for a scholarship from other countries."

International Exposure

For these workers going to other countries was crucial. One student explained that though workers can do "most of the courses . . . in our country, . . . we need to go to a place outside of the country . . . because of the element of exposure to what is happening outside." Employees are "encouraged to go out [because their work involves] contact with a lot of international organizations, [and thus employers] . . . want people to have much more exposure to what is happening outside our countries." Thus, nominated by their employer, Ghanaians go to China seeking exposure, knowledge, and credentials for employment related purposes.

One student admitted that he was decidedly unmotivated to go to China but went out of a sense of duty:

> Well, as a matter of fact, I came without any enthusiasm. . . . In the first place, China is very far from our place, and it's difficult to leave your family behind for such a long time. . . . There is no possibility to go home before you end the course, and most of my [previous] traveling outside the country had been short courses, seminars . . . two months, one month. . . . So, I was obliged to come. That's why I came.

Although this particular student felt obliged to go to China for employment related reasons, others went willingly, motivated by personal factors.

PART FOUR: PERSONAL MOTIVATIONS

On the questionnaire, four factors centered around potential areas of Personal Motivations in relation to family, educational authorities, desire to see the world, and desire to know China. The following table reveals these findings.

TABLE 6.4
Personal Motivations
Frequency (FQ) and Percentage (%) of Respondents who indicated the Level of Importance of Each Factor

	Very Important		Important		Not Applicable		Unimportant		Very Unimportant		No Response		Totals	
	FQ	%	FQ	%	FQ	%	FQ	%	FQ	%	FQ	%	FQ	%
10. Relatives and friends advised me to study abroad.	20	15	33	24.8	32	24.1	7	5.3	26	19.5	15	11.3	133	100
11. Educational authorities advised me to study abroad.	16	12	22	16.5	33	24.8	16	12	27	20.3	19	14.3	133	100
12. I wanted a chance to see the world	42	31.6	41	30.8	15	11.3	10	7.5	13	9.8	12	9	133	100
13. I wanted to get to know Chinese people and their customs	20	15	39	29.3	22	16.5	15	11.3	25	18.8	12	9	133	100

Advice from Relatives and Friends

As Table 6.4 indicates, 39.8 % (53) of the students reported that *advice from relatives and friends to study abroad* was 'a very important' and 'important' motivating factor. Two students acknowledged the influence of their peers. A Cameroonian wrote, "My friend who preceded me had left me a good impression by the mail, and the pictures that he sent . . . stimulated me a lot." And a man from Gabon paid tribute to "the contribution of friends and acquaintances" in his decision. In terms of family, some students wrote about their motivation to assert their independence. A student from Uganda revealed that he "felt . . . a chance to be independent from the influence and constant assistance of my parents." A student from Mali wrote that he went to China for, "the preparation of my future life in a liberal way and without parents' pressure." Finally, one man from Burundi noted in the margins that going to China was a "personal decision" that his family "were against" (Notes from five surveys).

On the other hand, most comments suggested that students with strong familial support had family with diplomatic ties to China. For example, one woman from Sierra Leone wrote:

> I was really convinced by my uncle who was the then Ambassador to China. He emphasized on the speaking of another international language. Since China was opening and not [many] people in Africa speak

Chinese, . . . I was told to take this opportunity to study in China since
he was there. I was grateful for that. (Survey Note)

A woman from Lesotho also wrote about her family ties to China. She
stated, "The first reason for me to come to China was to accompany my
Mother on a diplomatic mission. And of course, when I got here I was
offered a scholarship which was very important to me because I figured I
would still be in my country . . . [had I] not got the opportunity to study
abroad." Likewise a man from Equatorial Guinea wrote, "When my
brother was the Ambassador . . . in China, . . . he found the bursary."
Another Guinean also noted the influence of his family. He revealed, "I was
in China on an invitation for holidays. Once in China, my father asked me
to stay and study in China. So I stayed because he wanted me to do my
studies in China." Finally, a student from Madagascar also acknowledged
the influence of his mother and father. He wrote, "My parents know
Chinese, so I have to know Chinese and know the Chinese [people] and
China" (Notes from five surveys).

Advice from Educational Authorities

The *advice of educational authorities* had less influence than family and
friends. A total of   24.8% (33) cited the *advice of educational authorities*
as 'not applicable' while 32.3% (43) cited this as 'unimportant' and 'very
unimportant'. However, one student from Burundi did attribute his motivation to a high school math teacher.

I was in the last year of secondary school, [and] I was getting ready to
take the State Exam to go to the University. My Math professor
advised me to take an exam at the Chinese Embassy in Burundi which
was designed to select candidates to go to school in China. I went to
take that exam, and I passed it. That was an important factor that
made me go to school in China. (Survey Note)

Chance to See the World

Beyond teachers, friends, and family, however, the *chance to see the world*
was the most often cited motivator, second only to getting a scholarship. A
total of 62.4% (83) rated this factor in the top two categories of importance. Having travelled extensively with his diplomatic family, one Zaïrian
wrote that he was motivated because "Asia was one of the continents that
I hadn't yet visited." A Burundi man stated that "Besides [feeling] proud
for being out to do my studies, . . . [I wanted] . . . first, [to] see the foreign
world, the other people, and their cultures [and] second, to open up to the
outside world." Another Burundian was also motivated to become "open
to the outside world," adding "at home, we know what is going on in the
world, but Asia remains a difficult continent to know [because of] language, culture, etc." (Notes from three surveys).

Chinese People and their Customs

Some students declared a desire to "discover another world" while others expressed a more specific desire *to know Chinese people and their customs.* Though more students rated this factor as 'very unimportant' than 'very important', 18.8% (25) and 15.0% (20) respectively, a total of 44.3% (59) students ranked this as 'very important' and 'important'. Moreover, most written comments tended to relate to this factor.

Curiosity Factor

A few students wrote about "the curiosity factor," as expressed by one Gabonese. He added that "China is a country that is very far and not very familiar. It was then necessary to go to that country, to appreciate those mosaics, those traditions, etc." A Tanzanian woman was also curious, "to get to know Chinese people and their customs." Likewise a student from Guinea was motivated "especially [by] the discovery of a people so far away, strange, and closed." For a Cameroonian woman, "The Chinese people [were] an enigma. . . . I had a desire to know their lifestyle and how they react with foreigners." Finally, a student from Congo simply wanted "to learn about the Chinese" while a student from Benin wrote that he "loved China [and] . . . wanted to know [the people]" (Notes from five surveys).

Chinese Language

Beyond general curiosity and a desire to know more, a number of students commented specifically that they were motivated to learn the Chinese language. A man from Mauritius was motivated by the opportunity of "knowing to speak Mandarin" as was a student from Gabon who wrote that "besides travelling outside of my country, I [wanted to] speak Chinese and know Chinese kung-fu." Some students felt that learning the Chinese language would distinguish them. A student from Congo chose "China . . . to be special . . . because the people in my country do not speak Chinese." A woman from Botswana also felt motivated to distinguish herself. She wrote, "I was always interested in Asia . . . I thought it would be nice to be able to write and speak Chinese. I don't know anyone in Botswana who does." Finally, a woman from Lesotho was also interested in the language but specifically in writing. She noted, "Above all what excited me was that studying in China adds on something else that not everyone who studies abroad can get — learning or knowing how to write characters" (Notes from five surveys).

Chinese Culture

Learning the language is, of course, an entree into the culture, and many students were motivated to go to China because of its rich cultural history. A student from Sierra Leone noted that "China is one of the biggest countries and the most populated in the world with diverse cultures, traditions, norms, and values which are worth knowing." A student from Kenya reiterated this feeling, "China is the most populous nation in the world and has a rich cultural history." A third student also supported this view, "I had the choice between the former USSR and China, so I chose the last because I thought China is very rich culturally" (Notes from three surveys).

Other Cultures

Students wanted to know China but also other cultures. A student from Ghana was motivated by the "the opportunity to interact with people from other cultures." Likewise, a Sierra Leonean was motivated "to be able to appreciate the values in different customs and traditions, especially those of other foreign students." In addition, a student from Mali wanted "the chance to meet several races and groups and know better their problems, their culture, their way of thinking and reasoning" (Notes from three surveys).

PART FIVE: OTHER MOTIVATING FACTORS

Beyond the four areas of potential motivation I had proposed, Financial, Academic, Employment, and Personal, students offered many other factors in the open-ended section. These comments could generally be organized into five additional areas of influence: governmental, political, economical, developmental, and technical.

Governmental Factors

Students were motivated both by the Chinese government and their home government. A woman from Namibia noted that "the Chinese embassy in my country" was a motivating force while another emphasized, with capital letters and three exclamation marks, that he came to China because of "no other factors than the false impression given to me about the life in China by their embassy in Sierra Leone." While the Chinese embassies in these countries were influential, the home governments of many countries were also cited as a factor. A student from Ghana was motivated by "the factor that I did not know the conditions here and my government had spent money in bringing me to China." A man from Burundi wrote that "the main reason" he came to China "was because I was told that . . . they need people who speak Chinese . . . in the embassy of Burundi or in the field of cooperation between Burundi and China in general." A student from Sudan was also influenced by his government. He wrote, "The only motivating factor . . . [was the] . . . agreement between China and my country [for] an academic exchange. Hence, I am here under this umbrella. That is, I am the only one here for the agreement of country to country [relations]" (Notes from five surveys). While these students reported that they came to China out of a sense of duty towards their governments, other students came to China to avoid governmental instability and political uncertainty in their countries.

Political Factors

Four Rwandan students detailed how civil strife affected their educational opportunities. By way of motivation, one explained, "In my country, there was no stability, otherwise I would have continued in the university in my country." Another student also searching for stability added, "I am in China for the second time, not because I wanted it [but] the events in my country made me come, in order to find some peace. Really, I had a lot of disappointment in my life (war), maybe that is why I am not very touched by the world." Another student noted that in his country "There are not enough facilities to enable people to study well, even professors and teachers are not available since some were killed during the genocide that took place in Rwanda in 1994." Finally, one student explained the situation further:

> In my home country, there is a lot of trouble and for that reason, the university there has academic years without school. To finish an architecture degree there, it could easily take you 7 years instead of 5, and that is not encouraging at all. In my country, there is a lot of uncertainty. (Survey Note)

A number of Burundians expressed similar motivations. One wrote that, "The situation in my country (security wise) was deteriorating, and coming to China was a way to help myself stay away from it." A country mate added, "It's mostly the current political situation in my home country (civil war) that made me decide to come here." A third student was also motivated to go "because my country is going through a hard time. Also, I thought China was a country where I could have the means necessary to finish my studies, which is exactly the opposite for my country of origin." In addition to Rwandans and Burundians, a student from Mali was motivated to avoid civil "perturbations (strikes from time to time)," and one student from Congo wrote that, "The essential motivation was that I wanted to go out of my country," though he did not specify the exact reasons. Finally, another student from Mali seemed to summarize this issue:

> Now, many young Africans want to leave their respective countries for political reasons (civil war, instability). They want to continue their studies in other countries where there is peace. So the Chinese scholarship is welcome. (Survey Note)

Economical Factors

While some student felt compelled to leave their own countries, others were drawn to China because of the "economic factor," as one Gabonese put it, adding "China is now a growing economic power in the world." An

Ugandan also commented that "China [has] the fastest growing economy in the world, [and] this made me realize that they could have something to offer." Another student noted China's "fastest growing economy" and [hoped to] learn their magic formula." The same student also wanted to "get to know how the government is able to govern a gigantic population." These students clearly felt that China had much to offer beyond the classroom.

Developmental Factors

Many students were motivated to go to China in order to help their own countries develop. Some perceived that "China was sort of a lab for the developing world," as one Burundian wrote. Others linked China's development with the development of their own country. A Zambian stated, "China is a developing country, . . . and coming from a developing country I reckon I would be well exposed to actual experiences of various developments going on in China which will be very relevant in my home country." A student from Sierra Leone expressed similar sentiments. He wrote, "The developmental status in both Sierra Leone and China are very similar in nature, but China is a little ahead . . . in advancement of Science and Technology. He added that "Training in the appropriate technologies would enhance transfers of useful knowledge for the benefit of the [Sierra Leonean] community."

Technological Factors

In the interviews, the two Post Graduate students elaborated more fully on these factors of appropriate technologies, transfer, community benefit, and development. One student stated:

> I'm here purposely because my country is trying to develop and then also because [of] this concept of appropriate technology. . . . [This is] .
> . . one of the motivating factors that brought me here . . . We know back home that the Chinese have some of the now call appropriate technologies, . . . so I came here purposely to learn some of these basic technologies, what we call appropriate technologies, from the grass roots. (Post Graduate A)

The second interviewee expressed almost identical motivations:

> I was thinking that I would have a first hand opportunity to see really what they are doing here because back at home we usually think that China is a Third World country, a developing country, so maybe their technology, the level of technology, should be more applicable to ours than the West. So, we were really interested in seeing what they are doing here [and] their level. (Post Graduate B)

The same student continued as he defined his sense of appropriate technologies:

> We think that the technology that is more appropriate to a Third World country would be one that is being used in a Third World country. . . . We get our imports of machinery and high tech from the West. . . . They are expensive, and when they break down it is very difficult to replace parts because you have to import everything. So the idea is that if you get technology that is relatively simple, it will be more applicable to our stage of development right now.

> So the idea [is] that, you know, South-South. . . . The Chinese are a developing country, and they are also coming up in development right now. It is interesting to come see what they are doing and then see whether or not you can apply that to our situation back at home. So that was a real important thing, one real important consideration for coming here. (Post Graduate B)

SUMMARY OF MOTIVATIONAL FACTORS

In summary, the most often cited motivating factor was *obtaining a scholarship*. The majority of students, 82.8% (110), reported that *at the time of [their] decision to leave [their] country to study in China*, obtaining a scholarship was a 'very important' and 'important' consideration. Students indicated that without the scholarship all other factors would be secondary. *Getting a chance to see the world* was the second most often cited motivating factor, rated 'very important' and 'important' by 62.4% (83), and *China's facilities and course offerings* as the third most often cited motivator, rated 'very important' and 'important' by 54.9% (73). Employment prospects related to a higher degree, though not necessarily a higher degree from China, were also important motivating factors, especially in the case of Ghanaian students. Written comments from the open-ended question which closed this section supported these statistical findings and added to the discussion by raising other sources of motivation, such as governmental, political, economical, developmental, and technological factors.

Finally, it is worth noting a few comments which defied specific classification. One woman from Kenya regarded studying in China as a personal challenge. She wrote, "I had actually never considered studying anywhere else apart from in my country let alone studying in China. So the opportunity to study in China was sort of like a challenge, and so I decided it would be worth a try." A student from Uganda who cited all factors as 'very unimportant' explained, "Whenever there is a place or a chance to tap knowledge from please do. Never mind whether it's what or where, what you need is knowledge." While he was motivated to seek knowledge, a returning student from Cameroon was motivated by the "possibilities of doing business [and] mak[ing] contacts for creating small and medium enterprises." Finally, one Kenyan was motivated by religion. Quoting the Bible, he wrote, "My Christian faith requires me to 'go and make disciples of all nations'. . . . People have to hear the love of God and not bound in sin and immoral practices." He ends by motivating others to "read Matthew 28:16-20."

Notes

.[1] This section of the questionnaire was largely adopted from Mickle, M. (1984). *The Cross Cultural Adaptation of Hong Kong students at two Ontario universities.* Unpublished doctoral dissertation, University of Toronto, Ontario. The four specific themes of motivation, Financial, Academic, Employment, and Personal, emerged from a series of questions designed to touch upon all of the potential factors that influenced Africans to study in China.

.[2] Nationality of questionnaire respondents has been depicted for context.

CHAPTER SEVEN

Issues

This chapter focuses on issues that foreign students commonly face. In Section Three of the questionnaire, respondents were asked to rank, on a scale of 1 to 5, their level of agreement and disagreement on fourteen statements pertaining to student life. Table 7.1 presents these findings. This section closed with an open-ended question that enabled students to raise issues I had not.

PART ONE: LEVEL OF AGREEMENT AND DISAGREEMENT ON ISSUE STATEMENTS

Despite *often* feeling *homesick* (62.4 % (83) 'strongly agree[d]' and 'agree[d]') and *often* feeling *lonely* (37.6% (50) 'strongly agree[d]' and 'agree[d]'), the majority of students, 53.4% (71), 'agree[d]' to some extent that they *enjoy living on campus.* Only 28.6% (38) expressed some tendency to *suffer from occasional depression.* In the margins, one student wrote that he was homesick "only to visit my family and my friends," and another wrote that he was homesick "not very often but sometimes when I feel unhappy."

Approximately one third of the students, 33.1% (44), 'agree[d]' that their *Chinese classmates are friendly and helpful,* although four noted that this statement was not applicable as they did not have any Chinese classmates. The same percentage of students, 32.3% (43), 'agree[d]' that they enjoyed *close personal relationships* with both Chinese men and Chinese women, and 42.9% (57)'agree[d]' that they were *able to enjoy contact with fellow foreign students.* The majority of students, 82.7 % (110), 'strongly agree[d]' and 'agree[d]' that they had *adequate educational preparation at home* "but without the fluency in Chinese language," as one student pointed out. The majority of students, 62.4% (83), also 'strongly agree[d]' and 'agree[d]' that they felt *highly motivated to do well in school.* One student, however, felt that his motivation had been thwarted, "Naturally I am a very motivated person, but the education system here is not very stimulating to my personal experience. Maybe it is because of the field of my studies (International Politics) needed more debates and personal views" (Survey note).

The majority of students, 59.4% (79), 'strongly agree[d]' and 'agree[d]' that they were *concerned about racial discrimination,* though one student noted he had "been able to get over that aspect." Most responses, 39.1% (52), were neutral when asked about the *favourable attitude of the local people about their country,* but more, 33.8% (55) to 23.3% (31), tended to 'disagree' than 'agree' that the local attitude was *favourable.* A slightly higher number of students tended to 'agree', 33.9% (45), rather than 'disagree', 32.3% (43), that their behaviour was *often misunderstood* by the local people. In the margins, one student attributed these misunderstandings to "the big gap between our cultures," and another explained, "more often I have to readjust my behaviour to the Chinese counterpart." This notion of readjustment was echoed by another who wrote, "After six years in China I know what I can do." For the final issue, a large majority, 60.1% (80), 'disagree[d]' and 'strongly disagree[d]' that *the faculty [had] a reasonable knowledge of their country.*

Most respondents, 44.4% (59), 'agree[d]' that this list presented *a complete picture of the most important issues facing foreign students in China.* One Guinean wrote, "I agree entirely that the above list presents complete images of the aspects with which us, the foreigners, are confronted with in China." At the same time, 40.6% (54) 'disagree[d]' and felt the list was not complete. Many of those who felt the list was incomplete brought new issues to my attention. However, even those who indicated the list was complete chose to elaborate on the issues raised.

TABLE 7.1

Issues

Level of Agreement and Disagreement on Issue Statements by Frequency
(FQ) and Percentage (%)

	Strongly Agree		Agree		Neither agree nor disagree		Disagree		Strongly Disagree		None Response	
	FQ	%	FQ	%	FQ	%	FQ	%	FQ	%	FQ	%
1. I enjoy living on campus.	29	21.8	42	31.6	30	22.6	15	11.3	13	9.8	4	3.0
2. My Chinese classmates are friendly and helpful.	20	15.0	44	33.1	43	32.3	11	8.3	6	4.5	9	6.8
3. I am often homesick.	46	34.6	37	27.8	25	18.8	10	7.5	6	4.5	9	6.8
4. I am able to enjoy contact with fellow foreign students.	56	42.1	57	42.9	12	9.0	3	2.3	1	0.8	4	3.0
5. I am able to enjoy close personal relationships with Chinese men.	19	14.3	43	32.3	31	23.3	16	12.0	20	15.0	4	3.0
6. I am able to enjoy close personal relationships with Chinese women.	18	13.5	43	32.3	43	32.3	12	9.0	12	9.0	5	3.8
7. I am concerned about racial discrimination.	44	33.1	35	26.3	22	16.5	14	10.5	11	8.3	7	5.3
8. I suffer from occasional depression.	10	7.5	28	21.1	32	24.1	20	15.0	37	27.8	6	4.5
9. I am often lonely in China.	15	11.3	35	26.3	30	22.6	25	18.8	21	15.8	7	5.3
10. I feel highly motivated to do well in school.	43	32.3	40	30.1	25	18.8	11	8.3	4	3.0	10	7.5
11. I feel I had adequate educational preparation at home.	59	44.4	51	38.3	10	7.5	3	2.3	3	2.3	7	5.3
12. The attitude of local people towards my country is favourable.	8	6.0	23	17.3	52	39.1	22	16.5	23	17.3	5	3.8
13. My behaviour is often misunderstood by the local people.	17	12.8	28	21.1	39	29.3	26	19.5	17	12.8	6	4.5
14. The faculty has a reasonable knowledge of my country.	1	0.8	18	13.5	27	20.3	31	23.3	49	36.8	7	5.3

PART TWO: WRITTEN COMMENTS ON THE QUESTION-NAIRE

Students' comments frequently related to financial, language, academic, and social issues. I have reported these contributions in the sections which focus specifically on those topics. Issues that I did not account for in this particular section nor in any other part of the questionnaire were brought to my attention. In just a single word or two, a few students noted the weather, food, health services, living quarters, and general living conditions on campus. One student brought up issues regarding family life. In a rhetorical style he asked:

> For the married students, are there adequate structures on the campus to welcome their families? How about the education of their children if they are in China with their parents? If you are married to a Chinese person, is the connection of the cultures possible? (Survey Note)

Race

Beyond these notations, the number and depth of written comments on all of the questionnaires overwhelming pertained to race, discrimination, and bias. Many students even attached a separate piece of paper to communicate that "The most important issue facing foreign students is the question of racial discrimination and the class consciousness of the local people," as expressed by one Ghanaian. A Burundian specified this point, "The Chinese have the bad habit of looking down on black Africans." "Even teachers," according to one Ethiopian, "don't believe blacks can be good enough." A Benin man observed that, "The Chinese unconsciously think of an African as someone they should set straight." Another student simply noted, "They are critics of black skin." A Cameroonian woman elaborated, "The Chinese present a masked xenophobia towards foreigners, especially the Africans, therefore, they pretend to love us, but in reality, they despise us" (Notes from six surveys). As a result of these attitudes, one Namibian woman declared:

> African students are seriously molested and embarrassed in China. They [Chinese people] always tell their fellow Chinese that Africans are dirty, and that's why we are black. We sleep on trees and don't have food to eat. Because of these ugly words, Chinese class Africans as slaves and [feel] their so-called assistance is a big relief to all Africans in China. They always look [at] Africans with contempt, hatred, and malice. (Survey note)

Students linked how they are perceived individually to how the entire African continent is understood. "Our continent is the most damned according to their conception," wrote one Burundian. An Ugandan added that the "Chinese simply associate Africa and any country in it with disaster, poverty, wars, and all the bad things you can imagine." Another

Burundian asserted that "Regardless of what is meant in political speeches by Chinese officials in Africa, . . . when you are from a small country, they don't respect you anymore, and . . . [if you are] from a poor country, then [you are] not an important person." And yet another student acknowledged these same biases when he wrote, "Without denying the slow development of Africa, the Chinese seem to look down on Africa. As a proof of that, just look at the rare television programs. For them, all the bad things are African, nothing good comes from there." One student seemed to summarize these sentiments when he wrote, "If you look at the whole thing, our feeling, inside is that we are/I am not satisfied by these attitudes of many Chinese [who] look down on African students as [if] we came from nowhere"(Notes from five surveys).

Misinformation

Students largely attributed these 'attitudes' to "a lack of knowledge of the African realities." This ignorance, according to students, is because as a society "the Chinese are not open to the outside world, and they are taught things totally contrary to Africa." Again and again, this sentiment about the local population being "ill informed" and deliberately "misinformed" about African realities emerged. One student noted that "African students . . . are maltreated by the lack of information," and to a Malian this 'lack of information' explains why "there is always confrontation between Africans and Chinese."[1] In short, students recognized that "The existence of several bad prejudices on our countries of origin and ourselves . . . makes the connection between Chinese and foreigners difficult" (Notes from ten surveys).

PART THREE: THE INTERVIEWS

The nature and impact of this "difficult connection" became clearer in the interviews. The following answer to my question about *essential elements* typifies students' response:

> I think an essential part, you know, you should try to focus on how the Chinese people relate to foreigners in general, and more specifically to black Africans. Right, no matter your social status, the fact that you are black, you know, you are being here, how should I say, looked down upon. (Post Graduate A)

Daily Frustrations

In the interview with the Four Graduates, they also spoke about being "looked down upon" as the very first *essential element* they chose to raise were the "daily frustrations";[2] that is, the daily frustrations of walking down the street, talking the bus, and walking into an office. In these daily situations, students reported that they are insulted, shunned, and feared as a matter of course.

Insulted
One student described walking down the street and the verbal taunts that ensue:

> Every day, everywhere I go, [I] get the same thing. Someone is calling, 'hey black devil,3 . . . you are dirty, you are poor, you have nothing. . . . Black devil, black stupid, you are. . . .' Whatever — all dirty words. You leave one corner and then the next one. . . . Let's say you're walking a hundred meters . . . in between, you can get the same thing ten times, eleven times, in one hundred meters. (The Four Graduates)

One Embassy Counsellor also brought up the severity of these daily comments encountered in the streets. He concurred that "They do speak things like that":

> People in the street will say, . . . 'You come here [because] you need assistance, but you can't give anything. . . . You come for something, but you are not bringing in anything. . . . Black stupid, black poor, black strange, black very ugly, AIDS. . . .' Things like that. . . . Black — any negative adjective you can put on it. (Embassy Counsellor B)

The Counsellor explained that most "don't say it straight." In other words, these "negative adjectives" are usually delivered by people who assume that the students don't understand Chinese. However, "99, 98 percent of Africans here, they [do] understand the language. So, you are in . . . a shop, for example. They can be talking, not talking to you, but you hear what they say." The Counsellor added that even at times when students may not understand the language, they can understand the message:

> Even when you don't understand the language, for example, if you go to Shanghai [where] they speak Shanghai [dialect], you don't understand. But when somebody's talking . . . you can read it and see it. Yes. When it's a question of admiration, you can also see it, if the mind is really appreciating. Yes, you feel it. (Embassy Counsellor B)

Shunned
As the Counsellor pointed out, language is not a sole indicator because students can "read" and "see" people's reaction. This type of nonverbal reaction became evident when students described another daily frustration: taking the bus. For these students, taking the bus had become "a very strong issue". One student explained:

> [Despite over crowdedness] people always stand in the bus when there is a seat next to me. I do think, what's wrong with me? Do I smell? . . . I'm decent, you know. I don't see any [reason why] someone could not sit next to me. . . .
>
> Then, of course, if there is someone sitting on the window and I sit next to him or her, then he will, she will, feel like trapped. And then, if there

is another spare seat, it doesn't happen always, but it happens very, very often, she or he will move to another [seat], . . . but most [of the time], . . . maybe 70 percent, I will sit, like in two seats, and no one will come to sit next to me. (The Four Graduates)

For this student, when people refuse or vacate a seat next to him, they signal a clear gesture of impoliteness and disrespect:

If I sit next to someone and he just stands up, to just stand in the bus, obviously, it is elementarily impolite. You don't have to have a university degree to understand that. You've shown a lack of respect to that person. You cannot give him the luxury of sitting next to him. . . . It's just simple elementary impoliteness. And it happens always, very, very often. (The Four Graduates)

On many different levels these daily verbal and nonverbal affronts, though clear, leave the students mystified:

You know, it's really amazing when . . . I see someone digging a, a street and laughing at me, like poor, black devil. They insult me. I don't want be sarcastic, but I should feel sorry for him, but he feels sorry for me. He thinks I'm, I'm, I'm really meaningless because I'm a black devil. . . .

Somehow he's proud, he's proud of himself. He's allowed to be proud of himself, . . . more [than], . . . according to him, I'm allowed to be proud because I'm really a little thing. And this I cannot understand. I don't understand where, where all this pride comes from. And, I do not get [why] someone who has been digging the road and who has dirty clothes [and] I have a clean jacket or trousers, he cannot sit next to me. (The Four Graduates)

Feared

Perhaps even more perplexing than such "pride" is the reaction of fear. Another student in the group described the third daily frustration of entering an office building. He began with a specific example:

I had an appointment with the general manager of [a] company. . . . The secretary, . . . she's always someone who is educated, at minimum she's graduated from the university, . . . she came to open the door, [and] ahh!, she screams, she jumps . . . [as if she met] a tiger or a lion in the corridor. (The Four Graduates)

According to this man, "When they see me, . . . they are afraid, . . . they do fear me." The secretary's fearful recoils "shamed" this student, and it took him "maybe something like fifteen minutes or twenty minutes just to cool down, . . . to talk, saying everything confidentially, without any problems." Her response to him shook his confidence and raised many unanswerable questions in his mind:

I do think why, why . . . is she so scared? What does it mean? Why are
you acting like this? Am I an animal or what? Why, am I something
like, as a devil, or am I something that is coming from the other . . .
planets, or Mars or what? . . . I have a head, I have two eyes, I have a
nose. . . . I do ask myself, what does she or he think I am?

My answer is very strange. I cannot find a proper answer. . . . This
answer is, is full of alternatives. It's full of my feeling. What I feel is
very strong; it's even stronger than me, than myself.

The only thing different is the colour. . . . It's because of my colour,
they're not used to this. . . . So for me, he or she should know first
hand, I am a human being. (The Four Graduates)

Alienated

In three interviews, students spoke about this feeling of dropping "from the
skies" and feeling like "creatures" from other "planets." Students reported
that "People do not seem to understand why and how an African could
possibly be in China. Nothing can explain or can tell them why a black
man is walking in the Shanghai streets or Nanjing streets." On a few questionnaires, students also wrote about this sensation of being "some kind of
alien that's dropped down from Mars" and being perceived like an "Extra-
Terrestrial, like ET." One Zambian student wrote that these feelings inten-
sify during "certain times [when] the way people stare at me almost scares
me." In an interview, one woman reported she initially dismissed this staring in that it came from "just students." She understood that, "You know,
students can peer through the window and then hide from the teachers."
Yet she was disturbed when she realized, "Even the teachers don't care.
[They] can see [the students staring] but will not even bother to stop and
say, 'what are you doing here'?" Another woman from the same group
added that she could understand "if it was the first time they are seeing me
but these people downstairs [on campus] see me every day." This woman
differentiated between ignorance and familiarity:

If it [were] a matter of not seeing me before, [okay]. . . . But now it's a
matter of, you know, like, maybe just showing rudeness to me, or
[showing me] . . . I am not accepted in your land, or something like
that. Then it's . . . humiliating. It is really humiliating. (Group of Three)

A man in this same group also expressed feelings of humiliation and ensuing depression:

When you are walking down the streets, I mean, they start looking at
you. I mean, you feel so bad, you feel so depressed. I mean, they look
at you with so much amount of hatred, malice or what. I mean, it's no
good. I mean, when you walk, they come to touch your skin to see if
it's black. I mean, it's not good. (Group of Three)

In a different interview, one student reported that these same feelings of "humiliation," "depression," and "fear" become so "uncomfortable" that, at times, he found it difficult to go outdoors and especially difficult to conduct his research:

> People come and they stare and they stare, and you feel real[ly] uncomfortable. . . . They can really stare at you until you fall down. Yes, they can really stare at you . . . everywhere you go. So sometimes, [it is] very difficult to even go outside. . . . Even people on campus, they stare at you . . . [If you] have to go to the villages to collect . . . information, it's one hell of a problem. It's one hell of a problem. The whole village will come and surround you. Yes, that's really uncomfortable. (Post Graduate B)

On a questionnaire another student acknowledged this discomfort stating that "the staring . . . [can] really be awful" but rationalized the circumstances:

> For most of the people here, it's like the first time they're coming face to face with an African outside of the TV. . . and they don't know how to react or to handle the situation. (Survey Note)

All Across Society

From the rural village to the urban campus, from the street digger to the university professor, from a passenger on the bus to the professional office secretary, all students made a special point of emphasizing that the inability to know "how to react or handle the situation" came from all levels of society:

> Listen, there is one point I want to say. This is . . . my big point. [With] the Chinese, . . . educated or not, we still face the same reaction. . . . For example, [if] . . . someone who is digging the road [says] black devil . . . it's okay. . . . [I] . . . can be insulted by someone who is digging the road . . . because he doesn't understand this. But someone who is a university student . . . is supposed to be capable to know [better]. (The Four Graduates)

Questions

Students expressed further frustration about the type of questions that people "capable to know better" ask as a matter of course. In all but one interview with African participants, students and Embassy Counsellors both reported that they are asked "funny, funny" questions about their blood, their hair, the type of housing, languages, food, temperature, and conditions in their home countries. The nature and regularity of these questions added to the daily frustrations. As one student put it:

> It's like living in . . . the middle of a big ocean made of frustration.
> . . . There is little I can do. . . . I cannot stop on every street corner and
> say, yes, I eat at home, and yes, I have a roof, yes, yes, yes, yes, yes.
> (The Four Graduates)

In addition to this sense of frustration, students expressed incredulity and
sadness:

> What really makes me sad [is that] such kind of strange questions will
> be fine if you get them from people who have never been to university.
> But it's incredibly unbelievable when you have such questions from
> your . . . professor or your university classmate. (The Four Graduates)

Reflections

To these students, "the only answer . . . to this situation is that it is extreme
ignorance." Students contemplated the nature of this ignorance and linked
it to issues of language, economy, and development.

In two separate interviews, students pointed out that in the Chinese language 'Africa' translates to *Fei zhou*.[4] Students explained that *zhou* means
continent, and *fei* means nothing; thus, students concluded, "*Fei zhou*
means the continent of nothing!" Students protested this, "We were refusing the word *fei zhou* . . . [because] *fei* has a lot of negative [meanings]."
On the other hand, students pointed out that America translates to *mei guo*
which means beautiful country. Students asserted that the perception of the
country affects the perception of race. For example:

> When they see Michael Jordan playing basketball on T.V., he's not as
> black as I am. . . . Obviously, it doesn't mean that he is not black. He's
> black, but he is something else. (The Four Graduates)

The students explained what this "something else" is:

> Michael Jordan, he's black, but he's Western, very Western. I mean, he's
> American. He's a millionaire, billionaire, and he's from the U.S.A., the
> first developed country in this world. . . . He's black, . . . but he's not
> the same, black like me, from the bad lands. (The Four Graduates)

The "bad lands," as this student put it, does not refer to the whole African
continent. South Africa, as these students pointed out, garners respect
because it is economically developed:

> When they say, for example, that he comes from South Africa, then he
> goes, 'Oh, your country is developed, diamonds, uranium', and what-
> ever. They start talking about Mandela, start talking about everything,
> you see. But suppose I say I am coming from [my country], 'Huh, must
> be small country. I never heard about it. . . . ' They try to minimize me.
> (The Four Graduates)

These students attributed these attempts to "minimize" them to concepts held regarding the level of development and economic prosperity of their home nations:

> The people here they have a certain conception, a certain thinking about African people. And if you analyze this very well, this conception, how they are thinking about African people, how they [are] just mistreating African people, it's, it's just because of the problem of . . . maybe . . . this economy. [China] is just pointing [to] African people [thinking], they are very poor and dirty, they are worthless. (The Four Graduates)

Responses

Students reflected on how these attitudes and circumstances affect them:

> People shall think maybe it doesn't affect us. But when every day you are waking up or walking around, . . . you find that you have this kind of pressure. You know that . . . the people, they don't accept you very well. There is something very bad, like they feel they can just insult me, like 'black devil', anytime. It's okay, I think it's an insult. It doesn't change me, [but] it's not easy for me to accept. . . .

> Am I living with . . . some people who are considering me [as] an animal, or am I just an inferiority? I am human being, . . . but I am just maybe half to them. Sometimes we can have this kind of feeling because someone is, 'oh, how this creature arriving here?' (The Four Graduates)

Students spoke at length about dealing with this pressure. To cope and to avoid serious conflict, students pretend not to notice and pretend not to understand. As one post graduate put it:

> If you want to reply that will be another thing. [So] when people are staring at you and, of course, some of them even insult you, we just pretend that we don't understand. (Post Graduate B)

Other students also reported that they choose to feign ignorance, keep quiet, and walk away to "save face":

> If you don't want to get into trouble, . . . don't [want to] cause an arrest on the street, . . . you pretend you don't understand, and you walk your way. . . . And actually, it's better if you walk your way, and you keep it for yourself, then . . . you're sort of saving face. . . . [So] for most of the time, I say nothing. I pretend I don't understand. (The Four Graduates)

While students "save face" in public, they acknowledge that these situations leave a personal mark, "Of course, you have it. You've got it in your heart, . . . you've got it in your brain." Once in your heart and in your brain, the pressure builds, and students find it necessary to retreat:

> Getting this every day, this kind of problem, you find sometimes . . . you become aggressive, [but] . . . you must control. Maybe it's being used to these people, just insulting you every day, telling you this and that words, bad things. Sometimes, . . . you hide. I mean, you hide because you may be under pressure. (The Four Graduates)

Of course, in hiding, the pressure remains:

> You think about a lot of things. [You] feel, feel destroyed, . . . [like] you . . . don't belong. . . . You don't, you don't know exactly what you are. So you have to think about that. Perhaps, you are going to drink a beer⁵ . . . because if you . . . keep on thinking about that . . . you will never, never, never get the solution. You will never, never, never know why they do that. (The Four Graduates)

Although students feel they will "never know why" and will "never get a solution," they take some comfort in the consensus among their communities:

> Take any African student . . . who has been here in China, [and] you just ask one question, . . . 'how do you feel about the Chinese?' I can't say 100 percent, but it's at least 85 percent, you're going to find everyone is complaining about how they treat us. (Four Graduates)

One student concluded that "The general behaviour of the Chinese towards foreigners, especially blacks, . . . brings a lot of negative reaction from the students" (Post Graduate B). This consensus and negativity was acknowledged by all Embassy Counsellors.

PART FOUR: EMBASSY COUNSELLOR PERSPECTIVE

One Counsellor proposed that an *essential element* in the lives of African students was this overwhelmingly "negative reaction" from the students. In his capacity as Counsellor, he expressed his bewilderment and admitted:

> I don't know how to describe it. . . . Actually, it amazes me in a way. I ask myself why, why, why ? Four years, your first degree, three years, [another degree], nobody says anything positive, be it small, be it small. . . . I've never found anything positive when talking to students at all. Nothing, nothing, nothing. Everything, anything they talk about — Chinese food, Chinese behaviour, Chinese anything — really, you won't find anything good. . . . You won't find a single one talking anything good about China. (Embassy Counsellor A)

He added that the students' disposition "is not only negative . . . [but also] resent[ful]":

> The African students resent the Chinese. It's a resentment, it's a high resentment, you know. And they resent anything about them. They feel the Chinese . . . don't like them in the first place. They feel, they despise them. They feel they are not liked. They feel, the attitude is really negative, we can say resentful. (Embassy Counsellor A)

The same Counsellor explained that this negativity and resentment is because "somehow nobody amongst them has learned, they have never . . . known the root [of] the Chinese culture, [and thus they remain] as perplexed as everyone else." He added:

> You will find . . . you have two students who have been here for a number of years, you find them disagreeing on something about the Chinese culture because none of them knows. You find they are differing. They have lived here long, but they are differing . . . mainly because their guess is as good as anybody's who has just arrived that day. (Embassy Counsellor A)

The Counsellor revealed that some students, in China for ten years and fluent in the language, remain at same the level of cultural understanding as newcomers because they "have never been let in. . . . They have never been given the chance to understand the culture . . . [nor] participate in any of the Chinese way of life." He emphasized that the culture has never been "inculcated into them, [and] they have not lived within it practically [because] it has always been closed to them." Upon reflection, the Counsellor stressed what he "really wants to get at":

> The African has never got the chance to dig deep into the Chinese life . . . and really . . . get exactly what he came to do: to make the friendship here. They never manage to get it. . . . There has always been a barrier. (Embassy Counsellor A)

The nature of these barriers and their impact on personal relationships are further discussed in the next chapter on social contact.

Notes

.¹ The *Beijing Review* (1987 01 19, 1987 01 26), Cheung (1989), Delfs (1989), Scott (1986 06 19, 1986 06 26), Zhi (1989) are a few among many who reported on the Chinese-African campus confrontations in Nanjing, Wuhan, Shanghai, Beijing, and Tianjin between 1978 and early 1989. Crane (1994), Dikötter (1994), Sautman (1994), and Sullivan (1994) provide detailed examinations of these incidents and discuss their far reaching significance. These works are further discussed in the final chapter of this study.

.² Over thirty-five years ago, Hevi (1963: 183) expressed this exact sentiment in almost the exact words. He wrote: "Many of the things that are done to African students and of which they daily complain can be embodied in the concept of colour discrimination that pervades all social strata in China."

.³ Black devil, *hei gui*, is the most common epithet students reported they encounter. Sullivan (1994: 448) explains: "The meaning of 'devil' (gui) is derogatory in that it expresses various degrees of hostility toward foreigners, treating them as "non-humans" (i.e. without having the Chinese 'heart and mind')." In adding the word black *(hei)* to *gui*, the epithet takes on a racial meaning.

.⁴ Sautman (1994: 421) reports that after the anti-African outbreak in Nanjing (1988-89), African students also requested that the Chinese name for Africa be changed. Sautman, like the students in this study, points out that *feizhou*, a homophone for Africa, can be translated as "evil continent," whereas homophones of other countries, such as America (*meiguo* or "beautiful country") and England (*yingguo* or "brave country"), are by contrast complimentary.

.⁵ An Ethiopian student also brought up the issue of alcohol as a means of coping. On a questionnaire, he wrote, "Many students who have never tested [tasted] alcohol or any similar things end up being drunkards and irresponsible by the time they graduate due to a long disappointing life experience with Chinese. Chinese do not treat foreigners equally, they never treat their peoples equally either."

CHAPTER EIGHT

Social Contact

This chapter discusses the following three areas of social contact: sources of social companionship, sources of close companionship, and the frequency of social contact between African students and their Chinese hosts.

PART ONE: MAIN SOURCE OF COMPANY

The first item in Section Four of the questionnaire inquired about students' main source of company. As Table 8.1 highlights, 25.6 % (34) of students mainly associated with *other foreigners*; 15.8 % (21) mainly associated with *other Africans*; 9% (12) mainly associated with *people from their own country*; and 1.5 % (2) cited that their main source of companionship was with *people from China*.

Many students could not choose one main group, and so 46.7% (62) chose a combination of responses. Of this 46.7%, 24.1 % (32) of students checked two or more boxes including people from China while 22.6% (30) checked two or more categories of companionship excluding people from China. Thus, 74.4% (74) of African students indicated that mainly, they did not keep the company of local people.[1]

These statistical figures begin to paint a picture that is further supplemented by comments on the questionnaire and in the interviews. In fact, African students and Counsellors alike reported that issues of social contact and social relations were *essential elements* of their experience in China.

TABLE 8.1
Main Source of Companionship by Frequency and Percentage

Main Source of Companionship	Frequency	Percent
People from your own country	12	9.0
Other Africans	21	15.8
Other foreigners	34	25.6
People from China	2	1.5
Several groups excluding Chinese people[2]	30	22.6
Several groups including Chinese people	32	24.1
No Response	2	1.5
Totals	133	100

Barriers to Friendship

Typically, when students spoke about their friends they were referring to fellow foreigners because, for many, their "Chinese friends [were] very few" (Post Graduate B). This situation was contrary to the expectations of both Counsellors and students alike. As one Counsellor remarked, "One would expect if they came to China, they would have friends, Chinese, . . . they would become very close." He added, "After . . . three years time staying with the same students, in the same class, you would expect . . . that some friendship can be developed. . . . However, students just "don't see that coming" (Embassy Counsellor C). This reality perplexed many. Another Counsellor asked:

> Why, why, why, can you stay years in a country and you don't have a single friend? I always ask, . . . 'you guys, don't you have Chinese friends? Where are they? Where do they go?' Instead you have foreigners that you met here but not a single person in the country where you spent seven years, eight years. It's quite amazing.

> I often ask these fellows, 'you know, you have been here eight years, where are your friends? I want to meet them, and we will talk, and we get to know China'. And they say, 'ah, you are joking. You can never have a Chinese friend'. So that is the attitude really. You've been with them for seven or eight years, but . . . [you] can't make a friend. . . . Why? (Embassy Counsellor A)

Comments on the questionnaire began to answer these questions. Students reported that "There is a very clear barrier between the foreign students and the Chinese students. It is really difficult to make good friends among our comrades." One attributed this difficulty in making friends to "reasons of openness" and another to the fact that authorities "do not allow [local people] the chances of being in permanent contact with the foreigners." Other students suggested "the closure of the Chinese society" and the "interdiction of access to many places for foreigners (recreation, work)" separates people (Notes from five surveys).

Rules

In the student interviews, many expounded on the difficulty of friendship in light of these barriers. Students frequently spoke about the separate living and eating quarters for foreign students and the local population.³ One Counsellor revealed that these divisions first puzzled him:

> The first question [I had for students] when I arrived [in China],
> . . . what struck me, . . . you mean, you don't stay in the same
> dormitories as the Chinese?' They say, 'no'. Boys and boys? No.
> Girls and girls? No. We have our building, our own dining
> [hall]. (Embassy Counsellor A)

Students not only reflected upon their "own building" and "own dining hall,"⁴ but they also commented on the regulations for receiving visitors. Students pointed out that if they received a Chinese guest, the guest could only be from that university. Generally, no local guests from outside the university were permitted. Students reported that when entering the building, guests must check in with the guard, sign in a registry, note the time, and note the purpose of the visit. When leaving, guests must sign out and again note the time. Students described the environment in which guest must register, visitations are timed, and purposes are monitored as "frustrating," "restrictive," and "strange." Above all, such conditions were not conducive to developing close friendships and mutual understanding:⁵

> Foreign students are not allowed to have continual or perpetual
> contact with Chinese. They have to be observed, controlled, like
> you were [doing] something wrong. Obviously, they [the
> authorities] are waiting [for] something wrong to happen. So
> [guests], they have to be timed . . . and registered, and . . . so
> there's no really personal, private contact. And when I say pri-
> vate, it's like, just to be open up to each other, to talk freely, you
> know, call each other, or do things together, [like] cook together.
> (The Four Graduates)

Another student in this group concurred:

> We are living on a campus of thousands of Chinese students,
> and the only time you have to be together is when you are in
> class. But . . . when you are in class, you have no time to talk.
> And outside class, . . . you are put in a separate compound, and
> then, you are not allowed to get in touch. Sometimes, it really
> does not make things easier. (The Four Graduates)

For the students, these rules may not "make things easier," but one
Counsellor tried to explain his sense of these regulations from the Chinese
perspective:

> These are our rules, please respect them. . . . Among the rules,
> do not bring Chinese here, in the foreign building. . . . Don't
> mess around with our Chinese boys or girls. . . . Don't mix
> freely. (Embassy Counsellor A)

However such regulations are understood, the results remain. African students in China "can hardly make a friend. You actually live four years here,
but you wouldn't be having a friend, a Chinese friend" (Embassy
Counsellor A).

Students and Counsellors alike spoke about the risks for anyone who
may readily seek friendships and challenge the rules. According to another
Counsellor, "If these rules are not respected, and students don't stay put,
there are sanctions"(Embassy Counsellor B). He added that one responsi-
bility of the Foreign Affairs Office, (the *waiban*), was "to protect [and] to
prevent too much involvement of African students in Chinese [lives]." He
revealed that "Too much involvement in Chinese affairs can be one of the
reasons of bad relation between Chinese, the *waiban*, and the students."
According to the Counsellor "too much involvement" is "suspect," especially "too many friends, too many visitors, and ladies"(Embassy
Counsellor B).

PART TWO: CLOSE COMPANIONSHIP

On a number of questionnaires and in the interviews, students and
Counsellors broached the sensitive issue of relations with women. On the
questionnaires, students were more candid and revealed, "It is difficult to
feel at ease with a Chinese girlfriend because of the numerous rules of the
Chinese law, and also the Chinese society itself considers girls who go out
with Africans as being prostitutes." Another questionnaire comment reiterated this exact point, "Even though we are obliged to spend here in
China more than four years, an African student who entertains relations
with a Chinese woman, simple friendship or not, that means the practice
of prostitution in China."

For those who do have girlfriends or wives, students reported that both
parties endure difficulties.[6] One student illustrated this point with the following example:

> An African guy was married to a Chinese lady, and this is a real
> story. And they were walking down the street, and then some
> young Chinese, they started . . . mocking them and insulting
> them, insulting his wife. . . . And then the guy, because he was
> in China for such a long time, he understood everything, and he
> just grabbed the guy and . . . [said], . . . 'I'm legally married,

under Chinese law, my country's law, everything is legal. So, you accept it or not, she's my wife.' He took the guy to the police station and then the policeman said, 'Oh, okay we handled the case, blah, blah, blah.' Yeah, but of course, we know . . . this Chinese policeman . . . thinks the same. Because sometimes they [the police] say the same in our ears, and we hear. (The Four Graduates)

As a result of theses reactions, students reported that any relationships they may have are kept out of the public eye. In an interview, one man talked about the extent of this discretion:

You know, here in China, there are some African students . . . who have Chinese girlfriends. Why [do] those people, even today, [if they] want to go . . . shopping, they don't go with their girlfriend? Why? . . . If you have a Chinese girlfriend, [and] if you go with her, just to go one hundred meters . . . every Chinese you will meet will excite you and excite her. And those Chinese girls, they understand more than us, so sometimes you can see she wants to fight, she wants to cry because she is feeling very bad. (The Four Graduates)

The same student added that is it more than just a question of having a girlfriend because if an African student "has a European girl or American girl . . . or African girl, whatever, . . . together . . . they can go . . . do some shopping, eat outside, . . . [they] don't care." What is cared about, according to one Counsellor, is not so much interracial relations but relations specifically between Chinese and Africans. He stated that it is more than a question of "foreign relations . . . between a man and a woman." It is "a question of colour first":

For a Chinese here, it's more acceptable to see a white man going out with Chinese ladies. It's more acceptable to see a Chinese man going out with a white lady than seeing a Chinese and an African. When you, you are walking around, you are in the streets with a Chinese girl or man, people are going to turn out and look. There are people who still [do] not understand that such a relationship can exist because he's black. (Embassy Counsellor B)

While intimate relationships may be the most complex, the sense of divorcement between the local people and the students extends into everyday social contact.

PART THREE: FREQUENCY OF SOCIAL CONTACT

On a five-point scale from daily, weekly, monthly, yearly, to never, students were asked to indicate the degree of frequency of six statements pertaining to social contact with the host nation. In only two cases, the largest figure indicated "daily" contact: 40.6% (54) had 'daily' *social contact with Chinese people*, and 57.9% (77) *watch Chinese television programs* 'daily'. In all other cases, the largest number of students indicated that the frequency was 'never'. The majority of students, 55.6% (74) 'never' *have social contact with Chinese families*; 45.1% (60) 'never' *engage in athletic activities with Chinese people*; 36.8% (49) 'never' *read Chinese newspapers or magazines;* and the largest majority, 77.4% (103), 'never' *participate in Chinese student organizations*. Table 8.2 presents these findings.

TABLE 8.2

Frequency of Social Contact with Host Nation by Frequency (FQ) and Percentage (%)

	Daily		Weekly		Monthly		Yearly		Never		No Response		Totals	
	FQ	%	FQ	%	FQ	%	FQ	%	FQ	%	FQ	%	FQ	%
1. I have social contact with Chinese people.	54	40.6	28	21.1	15	11	9	6.8	24	18	3	2.3	133	100
2. I have social contact with Chinese families.	5	3.8	5	3.8	15	11	29	21.8	74	55.6	5	3.8	133	100
3. I engage in athletic activities with Chinese people.	14	10.5	20	15	14	11	17	12.8	60	45.1	8	6	133	100
4. I watch Chinese television programs.	77	57.9	28	21.1	10	7.5	5	3.8	12	9	1	0.8	133	100
5. I read Chinese newspapers and magazines.	28	21.1	32	24.1	10	7.5	11	8.3	49	36.8	3	2.3	133	100
6. I participate in Chinese student organizations.	4	3	5	3.8	8	6	9	6.8	103	77.4	4	3	133	100

To get a further sense of the dearth of social contact, one Counsellor suggested:

> Ask any of these students if they have ever been to a Chinese house. Ask them if the Chinese have ever invited them to go [out] together. Ask them if any of the teachers have ever invited them to their house to see how big their houses are, what is inside, how they live, how many children, and all that kind of stuff. (Embassy Counsellor A)

That "kind of stuff" was "part of what we have been coming to see" but in the end "don't get." The Counsellor believed students "don't get it" because the Chinese have a different view of their program. The Counsellor attempted to communicate the nature of the program from the Chinese perspective:

> We told you come. We'll give you tuition, accommodation, 500 yuan. . . . We'll give you what we promised, and please stay there. . . . Don't interfere with any internal matter of the Chinese. . . . We have given you what we promised, so lead a nice life with that. Keep your distance, put simply. (Embassy Counsellor A)

In other words, students may be welcomed to pursue academic endeavours but must confine their interests and pursuits to their studies:

> You must know what you ought to know [and] no more. I will teach you economics or engineering, but don't ask how a Chinese lives. I will teach you medicine, but please don't kiss a Chinese girl. I'll teach you veterinarianism [sic], but don't ask why [the] Chinese don't keep dogs. (Embassy Counsellor A)

Cultural Exchange

This Counsellor admitted that he found it a "little bit funny to call somebody from over there [in Africa] to come to China and when he reaches [China] say, you live here, you study here, but don't meet my people freely." Considering this situation further, he asked and answered an essential question:

> What is the objective of bringing you all the way from home [Africa], and then they call it a cultural exchange? If . . . the objective . . . was to make real friendship between the two countries, has it been achieved? No. I dare say no. (Embassy Counsellor A)

The Counsellor linked the degree of cultural interaction with the overall objective of the exchange and reflected upon why "the so called cultural exchange doesn't live up to its objective."

> Our students come to study the Chinese language, get a Chinese education, but they find they are limited in the interaction they have with the Chinese counterparts. They live apart. They are not allowed to mix freely. They don't visit each other freely. Actually, the social intercourse is so, so, so, limited, at the end of the day you wonder, . . . what the objective was all about. . . . It's quite amazing. So I link it with the objective of that exchange. The objective has not been achieved. It is shooting short of something. (Embassy Counsellor A)

Protectionism

The Counsellor "personally believe[s]" that the objective has not been achieved because the Chinese authorities "don't want that cultural interchange to be." He attributed this resistance to protectionism:

> They want to protect their children, their future citizens, from any other cultural influence they may have. . . . The Chinese want to resist the foreign influence. I mean, they want to protect their people from any other cultural influence. . . . They say they are opening up, but . . . I don't know. . . . It's pure protectionism, . . . protectionism against a foreigner. (Embassy Counsellor A)

Mutuality

According to the Counsellor, the "protectionism" goes one way. In other words, "They are protecting theirs against us, they are not protecting us against them. . . . They are protecting their people from being influenced by any foreigner whatsoever, not the other way around." He added that, "They don't want the Chinese to be influenced by the foreigner, [but] they want the foreigners to be influenced by the Chinese" (Embassy Counsellor A).

Like the Counsellor, students also raised the issue of protectionism and believed that this protectionism hindered the mutuality of the exchange. Students, on a personal and academic level, regretted this lack of mutuality:

> We have come to China first and foremost to learn and to acquire degrees and also to understand the Asians, like more specific[ally], the Chinese. Right, so to be able to do that, I think, there should be this interaction. To be able to understand the Chinese culture or the Chinese way of doing things, I should come into contact with the Chinese so that I will know exactly what it takes to be a friend to a Chinese. . . . See, for you to understand someone's culture, I think there should be interaction between the different backgrounds. So I think there should be enough interaction between Chinese students and foreign students for us to learn. You know, it's a mutual sort of thing, you learn from me, I learn from you. . . . But that's not the situation. (Post Graduate B)

In another interview, a student saw this lack of mutual exchange, learning, and understanding at the heart of many conflicts:

> If it has to be called cultural exchange, [they] have to make it in the full, the whole sense. I don't understand why this segregation, why? And this 'why' could be the answer to this cycle of

conflicts.. . . . If African students and Chinese students . . . are
fighting each other, it's because . . . this conflict is created, . . .
is built by the whole system. . . . We are sharing classrooms, but
we cannot share a social life. (The Four Graduates)

Conflicts between local and foreign students may be one result of this lack
of contact, but on an individual level, students also noted its impact. On a
questionnaire, one student from Guinea Conakry, now in China for over
ten years wrote, "Under any circumstances, I always feel a foreigner in
China." This feeling of being perpetually foreign was echoed in an interview:

I've been already in this country for already the last seven or
eight years. . . . I have been accustomed to the food, to the language, to everything, . . . but I don't belong here. . . . I asked
myself, when . . . can I feel like I have entered this culture? . . .
What will it take? . . . I don't know, but it's something I cannot
help because I cannot get used to it. . . . Every single day I wake
up, and . . . I feel like a stranger. (The Four Graduates)

A Balance

All Counsellors acknowledged that students "have not met adequate relations." Students feel "They have not been able to integrate within the society, and they feel they are not accepted here in China" (Embassy
Counsellor C). This same Counsellor suggested that though this lack of
integration and acceptance may "have an impact of regret, [students] manage to go around their education fine. Some even may use [the situation] to
their advantage because then [they] must achieve what [they] came for."
The Counsellor added that "In one way it means, well, I'm not meant for
here so no problem. . . . I'll get my education and go home." Students, having "now committed much of [their] energy [on going to China], concentrate on getting [their] degrees":

I have seen many of them successfully finishing their degrees
completely on time, and immediately they are told it's time to
leave and go home. . . . So they may not stop regretting, . . . but
they do not complain about it. . . . They have their degree in
their pocket, [and] they are grateful for it because, after all, it is
sponsorship. They are not spending out of their own pocket,
and [students] say, China has given us a scholarship, sponsored
our education, thank you. (Embassy Counsellor C)

Thus, according to this Counsellor, "there is a balance." Students "may be
dissatisfied with lack of integration . . . but not . . . with the content [of
their education] and the way they are being taught." In many ways, the statistical findings from the next chapter on Academic Experience support the
Counsellor's observations.

Notes

.[1] I calculated (1.5%) + (24.1%) = 25.6% keep the company of local people. Thus, 100 - 25.6 = 74.4% do not.

.[2] The last two options were not on the questionnaire but were added during the data analysis to accommodate student responses.

.[3] Generally speaking, in a Chinese university, all students and much of the staff live on campus. Students attend class, eat, and stay together in the same dormitories. Foreign students may attend the same classes, but they live and eat in separate quarters. The living conditions given to foreign students are fundamentally better than those of their Chinese colleagues. For example, most foreign quarters have daily hot water, heat in the winter, and more space. While students may enjoy these higher standards of living, some view these conditions as means to enforce segregation. More than thirty years ago, Chen (1965: 117-140) detailed the nature and consequences of this "privileged segregation enforced upon them," and Hevi (1963), throughout his text, commented on this situation in much the same way as students did in this study.

.[4] At one institution in Shanghai, students also spoke about "their own floor." They contend that until a violent protest a few years ago, foreign students were further segregated within their dormitory, by floor, according to their place of origin: Africans on one floor, Arabs on another, Koreans on another, Europeans on another, and so on.

.[5] Hevi (1963: 130-131) also noted that segregation bred mutual ignorance. He stated, "Through no fault of our own, we foreigners formed a segregated colony in the school, eating, studying, sleeping separately and even having separate entertainment, with the result that, though living in the capital of China, we knew extremely little about what the Chinese were really like."

.[6] Hevi (1963:131) reported that Chinese women who associated with African students were "packed off to prison or to the commune farms for hard labour, . . . their only crime being that they dared to make friends with Africans, contrary to Party's orders." Henry (1976) relates the experience of a Tanzanian student who after being arrested twice by the local militia "spent several nights in jail, in those underground tunnels, as penance . . . [for] attempts at seduction." Scott (1986) reported on a Liberian student who in 1985 "ended up alone in a jail cell for four days . . . [because of] . . . contact with a Chinese girl." At the time of this study, sanctions seemed to be primarily social not legal.

CHAPTER NINE

Academic Experience

This chapter explores students' academic experience in two parts. In Section Five of the questionnaire, the first set of questions asked students to evaluate nine areas of their academic experience. The second set of questions asked students about the relative ease or difficulty regarding twelve academic tasks.

PART ONE: EASE AND DIFFICULTY IN ACADEMIC TASKS

Overall, students did not report many difficulties in their studies. On a five-point scale from 'very easy' to 'very difficult', the largest number of students found the following four items 'very easy' and 'easy'. *Completing course work* was 'very easy' and 'easy' for 39.9% (53). *Using the library* was 'very easy' and 'easy' for 45.8% (61).[1] *Working in cooperation with classmates* was 'very easy' and 'easy' for 41.3% (55), though many students wrote that they "do not have Chinese classmates." Finally, *speaking in front of others* was also 'very easy' and 'easy' for 45.8% (61).

One Master's student acknowledged that his evaluation of the program was "highly influenced by the type of studies I did." He wrote "The only thing we did was follow the book without criticism." He encouraged professors to "allow debates [and] critical [thinking] . . . in class," and he added:

> It is very important to mention that as international politics students, we have never been given a chance to comment freely on international affairs. I was going to the same class with Chinese students, and I'm sure they have the same feeling. (Survey Note)

For all but one other item on this list, the largest number of students ranked them in the middle category finding them 'not easy but not difficult': *understanding lectures* 42.9% (57), *writing term papers and assignments* 45.9% (61), *establishing rapport with professors* 42.1% (56), *getting aca-*

demic advice 43.6% (58), *taking exams* 45.1% (60), *selecting courses* 39.1% (52). In regard to exams, one student noted that "conditions differ from professor to professor." In regard to courses, one student added, "There are not many choices in selecting courses." Another student pointed out that "As a Bachelor's student, I cannot choose the courses," and a graduate student found courses were "imposed regardless of . . . personal interest." For 34.6% (46) *taking notes in class* was 'not easy but not difficult,' but the same percentage also found note taking 'difficult' and 'very difficult'. One student noted that "The Chinese language is very difficult to write, [and thus] taking notes in class [is difficult]." Finally, the only item students rated more difficult than easy was *communicating with school authorities*. For 42.8% (57) of the students, *communicating with school authorities* was 'difficult' and 'very difficult'. Table 9.1 displays these findings.

TABLE 9.1

Degrees of Ease and Difficulty in Academic Tasks by Frequency (FQ) and Percentage (%)

	Very Easy		Easy		Not Easy but not Difficult		Difficult		Very Difficult		No Response		Totals	
	FQ	%	FQ	%	FQ	%	FQ	%	FQ	%	FQ	%	FQ	%
1. Understanding lectures	8	6.0	39	29.3	57	42.9	20	15.0	4	3.0	5	3.8	133	100
2. Writing term papers and assignments	7	5.3	41	30.8	61	45.9	15	11.3	2	1.5	7	5.3	133	100
3. Taking notes in class	10	7.5	28	21.1	46	34.6	36	27.1	10	7.5	3	2.3	133	100
4. Selecting courses	13	9.8	29	21.8	52	39.1	23	17.3	4	3.0	12	9.1	133	100
5. Communicating with school authorities	13	9.8	16	12.0	40	30.1	35	26.3	22	16.5	7	5.3	133	100
6. Using the library	18	13.5	43	32.3	44	33.1	22	16.5	2	1.5	4	3.0	133	100
7. Establishing rapport with professors	19	14.3	28	21.1	56	42.1	17	12.8	7	5.3	6	4.5	133	100
8. Speaking in front of others	22	16.5	39	29.3	46	34.6	13	9.8	6	4.5	7	5.3	133	100
9. Getting academic advice	14	10.5	26	19.5	58	43.6	24	18.0	7	5.3	4	3.0	133	100
10. Taking examinations	12	9.0	33	24.8	60	45.1	13	9.8	5	3.8	10	7.5	133	100
11. Completing course work	13	9.8	40	30.1	49	36.8	19	14.3	2	1.5	10	7.5	133	100
12. Working in cooperation with classmates	16	12.0	39	29.3	39	29.3	28	21.1	6	4.5	5	3.8	133	100

School Authorities, Students' Rights, and Respect

On the survey, many students discussed the difficulty of their relationship with school authorities. One student wrote that "Among the challenges with which the foreign students face [is] . . . an excessive control of our private life by the authorities." Another added that "Foreign students are highly cautious of the fact that the Chinese authorities are always keeping a suspicious eye on them." One student felt that in terms of "individual accomplishment, I could do much more in a country which has less police control [and more] freedom of action." Three others observed that "many basic rights, like the right to associate " and "the right to move," and the "right for students to freely worship God with each other" were not granted. Students linked issues of "respect" and "attitude" of the school authorities to the "confusing management of the material and spiritual rights." Part of this confusion lay in the fact that "universities are not governed by the same social law which makes it a challenge between brothers to see one law for one and not for the other" (Notes from eleven surveys).

In fact, "the way these university authorities . . . treat African students" was an *essential element* for one Embassy Counsellor. He began by explaining that a typical African university is unlike a Chinese university in that the entire academic community does not live within the same parameters. He stated, "At an African university, you might come to school [but live] far from your teachers." Beyond geographical distance, however, at an African university, the lecturers maintain a professional distance and evaluate students solely on academic matters. Students personal lives do not come into consideration:

> What he is going to consider is your paper, [your] participation
> in class, the way you act in class. But he's not suppose to know,
> he cannot know, at what time you sleep, . . . how many days
> you go [out], how many beers you have a day. No, no, no, no,
> no. He can see it on the way you work, he can judge [the work].
> (Embassy Counsellor B)

Yet at a Chinese university the relations between professors and students were "completely different." In China, the university community knows details of your life "they are not supposed to know. . . . [Even] the one who is working in the [cafeteria] is going to know [all about you]." In other words, "There is no complete separation . . . between life . . . and academic issues" (Embassy Counsellor B).

This Counsellor wanted to "insist on that [point] because it is so different from [African universities], and it is one main problem foreign students have." He spoke at length about the "way the *waiban* deals with foreign students . . . [and the] . . . relationship between teachers and those *waiban*." He explained how the *waiban* and the teachers "exchange information and

... influence each other, even influence your academic records " (Embassy Counsellor B).

The influence of administrative authorities on a students' academic records was specifically noted in one questionnaire: "The student's grade is often related to his rapport with a professor in management or the office of international students and not his academic competence." The Counsellor was even more explicit: "There are very good students who fail because of the *waiban*, and there are very bad students who pass because of the *waiban*. " He elaborated:

> There is no difference between academic [relations] and [relations with the] waiban. The waiban can ask academic teachers to let you pass. . . . Yes, and they can also say, please this man is not doing well, [do] not pass [him]. . . . The waiban can do it. . . . They can do it, and they do it. (Embassy Counsellor B)

In the following passage, the Counsellor stated his case by illustrating how poor students may pass, and good students may fail depending upon their relationships with the *waiban*:

> When you have good relations with the waiban, the waiban can say . . . to the professor, '. . . please . . . you know, [this student] is failing, but [he] is a very hard working student. Every night you see that man staying in his room, doing his best, reading all the time. . . . He's . . . a foreign student who is failing only because of the language. Please, if you could do something for him.' There is that.

> On the other hand, [take a] more intelligent [student], five or even ten points [ahead] but still under sixty percent, which is a failure. For that one, the waiban would say, 'do you know why he fails? All the time, he is having three beers. When others are going to study, he is going away for dancing. . . . ' And you fail. And it can be wrong. (Embassy Counselor B)

The Counsellor reported that students must be highly cautious of what they say and do, in and outside of the classroom:

> You don't say [to yourself], . . . what I'm doing, what I'm saying is for the teacher [because] you never know. After a hour it can be reported to anybody else. When you have a very small problem in your room, something is [broken], you don't say, this is to do with the *waiban*. No. It might be . . . put to academic authorities. . . . [Any] small problem . . . affect[s] your academic record. (Embassy Counselor B)

As a result, students "are very nervous [about] their social life [because]. . . you never know exactly . . . where you are, what you are doing, what you have to do here or there, . . . whether to pay attention. [You must] watch out, . . . anywhere you are, you [must] know what you are doing in everything, in everything" (Embassy Counsellor B). Students' academic experience was necessarily affected by this environment.

PART TWO: EVALUATING ACADEMIC EXPERIENCE IN CHINA

On a scale of one to five, from 'excellent' to 'not applicable', students evaluated their academic experience in nine areas. This data is displayed in Table 9.2.

TABLE 9.2
Evaluation of Academic Experience by Frequency (FQ) and Percentage (%)

	Excellent		Good		Fair		Poor		N/A		No Response		Totals	
	FQ	%	FQ	%	FQ	%	FQ	%	FQ	%	FQ	%	FQ	%
1. Orientation	6	4.5	54	40.6	35	26.3	25	18.8	6	4.5	7	5.3	133	100
2. Faculty assistance	14	10.5	45	33.8	48	36.1	19	14.3	3	2.3	4	3	133	100
3. Quality of instruction	15	11.3	50	37.6	52	39.1	13	9.8	1	0.8	2	1.5	133	100
4. Relevance of course content to home	20	15	50	37.6	37	27.8	6	4.5	14	10.5	6	4.5	133	100
5. Program design and requirements	11	8.3	56	42.1	40	30.1	16	12	3	2.3	7	5.3	133	100
6. Intellectual stimulation in general	14	10.5	54	40.6	38	28.6	19	14.3	3	2.3	5	3.8	133	100
7. Relevance of courses to future	28	21.1	49	36.8	38	28.6	5	3.8	7	5.3	6	4.5	133	100
8. Involvement in student	11	8.3	29	21.8	45	33.8	26	19.5	16	12	6	4.5	133	100
9. Overall satisfaction with the Chinese educational program	10	7.5	47	35.3	50	37.6	14	10.5	7	5.3	5	3.8	133	100

As the table indicates, the majority of students evaluated the following aspects of their academic experience as 'excellent' and 'good': *program design and requirements* 50.4% (67), *intellectual stimulation* 51.1% (68), *relevance of courses to [their] country* 52.6% (70), and the *relevance of courses to [their] future plans* 57.9% (77).

Orientation of international students was found to be 'excellent' and 'good' by 45.1% (60) of students, and the exact same percentage found it to be 'fair' and 'poor'. Likewise, 48.9% (65) found *the quality of instruction* to be 'excellent' and 'good' while the same percentage, 48.9% (65), evaluated it as 'fair' and 'poor'. One student felt that the instruction "could be better because sometimes the professors advance on the rhythm of the most intelligent student" while another student noted that he "appreciate[s] [that] the teachers are hard working" (Notes from two surveys).

The majority of students evaluated the following aspects of their academic experience as 'fair' and 'poor'. More than half, 53.3% (71), of the students indicated that *the level of involvement with student activities* was 'fair and poor'. One student wished this level of involvement was higher, and recommended that "foreign students [be] integrated into other school activities [such] as, cultural programmes, entertainment, and sports, out of which that are often sidelined." Finally, 50.4% (67) of students indicated that *faculty assistance* was 'fair and poor'. One student "suppose[d] the biggest obstacle is lack of personal touch with either the professors or Chinese colleagues . . . which . . . makes a difference in academic life. They don't seem willing to assist." Another student wrote that, "Our Chinese classmates seem to have more access to the lab material than us, this to say that our academic supervision is far from being very satisfactory" (Notes from four surveys).

Overall Satisfaction

As indicated, 42.8% (57) of students rated their *overall satisfaction with the Chinese educational program* as 'excellent' and 'good'. On the questionnaires no students specifically commented in this area, but in the interviews one student expressed satisfaction about learning traditional Chinese techniques for preserving fruit. He explained that in China when fruit falls off trees, they gather the fruit, take it to the kitchen, and preserve it. Yet in his country, fallen fruits are "thrown away . . . because they don't have that [preservation] technique." By studying the shelf life of mandarin oranges that he canned in a processing lab, he was learning traditional Chinese preservation techniques that were "scientifically based" but at the same time "very simple." He described the techniques as "home made . . . [requiring no] sophisticated equipment or machinery." He claimed, "All you need [is a] cooking pot, [and] you can turn your kitchen into a processing plant." Because fruit preservation was not widely practiced in his country, he looked forward to introducing these new techniques at home (Post Graduate B).

The same student was also studying traditional Chinese methods of storing vegetables. In his courses, he had been studying mature green tomatoes which were harvested, refrigerated, ripened, and then sold. Now he was preparing his Master's thesis on chilling injury. He explained that chilling injury occurs when tropical produce gets damaged while being stored in a cold environment. The damaged fruit has to be destroyed, and thus "injury causes a lot of losses." For his study, he was investigating a hot water method to pre-treat tomatoes to prevent or alleviate chilling injuries. He found that his thesis work "very useful," and felt he was learning "something new" because he had not studied "this hot water application on tomatoes before." Moreover, he was pleased because this method "worked quite well." He felt this technique could be applied at home because "people haven't been using that hot water method for tomatoes."

Thus, this student expressed satisfaction about learning traditional Chinese techniques for preserving fruit and storing vegetables. He reported that both techniques were new to him, useful, relevant, and directly applicable alone or as supplements to indigenous methods of his own country (Post Graduate B).

Unfortunately, this student's level of satisfaction was atypical. Although, as mentioned, 42.8% (57) of students rated their *overall satisfaction with Chinese educational program* as 'excellent' and 'good', a slightly larger percentage of students, 48.1% (64), rated their level of satisfaction as 'fair' and 'poor'. Among those who found their experience to be 'fair' was a Master's student from Ghana. He wrote, "China is not ready and ripe enough to accept foreigners, especially from Africa, as students to study in their institutions. To do this a lot of structures have to be put in place." Among those who found their experience to be 'poor' was a Doctoral candidate who wrote, "The academic program for foreigners is a total failure. I am quite disappointed. It seems like the Chinese people care less about us. They seem to undermine us." In the interviews, students chose to focus and elaborate on these feelings of disappointment and neglect.

> In the interviews, students reported "We, who come from Africa, . . . realize . . . there is something lacking." Students expressed that an essential element of their experience in China was a sense of unfulfilled expectations in the following three academic areas: program and course design, level of mutuality, and access to technology and industrial sites.

Program and Course Design

Students contended that "The universities don't have any well defined program for foreign students, especially in the post graduate area." One student qualified his remarks, stating he does "not know much about those who are reading their courses in Chinese, [but for] those of us who are reading our courses in English, . . . they don't have any specific program . . . for foreigners." A second student corroborated:

> I came here to do a course in food technology. I wanted to do some research in storage of fruits and vegetables, . . . [but] . . . when I came there was nothing. . . . All their programs are maybe designed to meet the [needs of] Chinese students, but foreigners, we don't have anything. Yes, they don't have anything for foreigners. (Post Graduate B)

Students reported that the universities "don't have a program for foreigners," and students must "scramble to find courses . . . just to get [the] minimum credits." And those minimum credit requirements "keep on changing . . . every year." Students stressed "it is very important . . . [because] those who are a year ahead of us are . . . are taking something different, meanwhile you are going to be awarded with the same [degree]" (Post Graduate A).

Moreover, students felt the universities should consider their background and experience. Students were "not very happy" because after they "scrambled," they found many of the courses to be "just [a] repetition [of] what [they had] done before." One student illustrated his point: I've been in the fruit sector . . . [for] seven years. After graduation,

> . . . I've had quite a lot of experience. . . . So at the graduate level, I was expecting that the course would . . . give me new ideas. [But] we talk about concepts in food technology . . . [that] I read about seven years ago. (Post Graduate A)

He recommended that in addition to formalizing academic matters, academic authorities should "look at the person's background, . . . look at the person's first degree, and . . . design a program that will make the [student] . . . capable of doing research" (Post Graduate A).

While these two students recommended that university authorities consider their personal backgrounds, another student urged university authorities to consider Africa. This student asked, "What conditions [do the Chinese authorities] have in the back of their mind[s] . . . for selecting courses . . . they are sending for scholarship . . . [to Africa]? What criteria [does] China use to offer courses for various African countries?" (Group of Three). In other words, he felt China offered course scholarships to African countries without adequate consideration of African conditions:

> When [China] advertises for scholarships, the courses for each offer have been designed in China. They don't [design] them to Africa. They want students to come and study this course, this course, this course, without taking into consideration what is there in Africa! When we came here, they gave us topics that are purely Chinese based. (Group of Three)

For his thesis, this student reported that he had the choice between two "purely Chinese based" topics: rapeseed or soya bean. Neither are found in his home country.

> We are in West Africa. We don't have any thing like rapeseed, oil seed. I mean, rapeseed, we don't have oil in my country. It's not there. It's purely a winter crop, and we don't have winters in my country. It's not important in [my country]. When [we] get here, they said we should do our thesis based on rapeseed or soybean. Soybean in my country is not important. The most important food in my country is rice! Potato, rice, cassava, sorgo, millet - these are most important things. (Group of Three)

Despite what is important in his country, he felt he had "no choice, no choice definitely." He added, "What I want counts for nothing." He felt incredulous about the situation and felt his time was being wasted:

> Can you imagine, like me, coming here doing this research in
> rapeseed? I mean, there [is] no rapeseed in [my country]. So
> what's the use? What's the advantage of that? It's just like wast-
> ing time. (Group of Three)

He continued that if China "really wanted to help African countries, as
they are claiming," China should ask questions and seek input. He recommended:

> [China] should ask the African countries which area of devel-
> opment they need most, . . . and from there offer scholarships.
> [Moreover], let the African countries design their own program
> and send students to come and study based on their program
> that is related to the development in their country. (Group of
> Three)

Finally, he asserted, "If you want to help someone, don't do something that
is purely in your own interest. Do it in interest of the other man's position"
(Group of Three).

Mutuality

Students did wish that the universities showed more interest in their position. In addition to lacking interest in their personal background and conditions of their country, students also felt that their schools lacked interest
in mutual academic exchange. One student illustrated this point by referring to two of his seminars. At the first seminar, he spoke about the role of
the national standard institutions and quality management. At his second
seminar, he spoke about his country's traditional approaches to food production, processing, and storage. He pointed out that the information in
both conference papers was "put together by his personal experiences in
the field [and as such would] not be found in any textbook or timetable,
either at the undergraduate or the graduate level." He emphasized that
"such knowledge and ideas can *only* be gotten from people who are in the
field." Yet at both seminars, his Chinese colleagues were "no where to be
found." As fellow graduate students, he expected "them to be there so that
at the end of the seminar, through the questions, and answers, suggestions,
. . . we can learn from [each other]."

   This student found the situation "not pleasant" and "strange" and
speculated on four possible reasons why, although he was "talking about
quality control systems [to] future quality control managers, . . . very few,
few, few [came to listen]." First, his colleagues may "not [be] interested [in]
new concepts and ideas that I am bringing from my home country."
Second, his colleagues may "think there is nothing that they can learn from
foreign guys that they don't already know." Third, lecturers "do not
encourage students." He compared how lecturers at home encourage students' interest and attendance:

Back home when a graduate has presented a seminar, the pub-
licity is quite enormous. Everybody gets to know. You go to a
library, it is there. And you [see], I'm interested in this topic
because maybe you want to go into that area. You always want
to learn from others. Maybe . . . the student has to . . . adver-
tise. You do that, and you expect the lecturers to disseminate the
information for graduate students. That aspect I don't think
they are doing well. . . . Maybe the department [does not] . . .
encourage that. (Post Graduate B)

Finally, this student linked the lack of encouragement and interest in academic exchange to a
lack of social exchange, "I don't know why, maybe they don't want to
interact with foreigners." In all he concludes, "So this mutuality, I learn
from you learn, you from me, that is not here. It is lacking" (Post Graduate
A).

Another student also highlighted his seminar experience to illustrate the
lack of exchange. In passing, he mentioned that no Chinese colleagues
attended his seminar either, but he focused on his professor. His professor
expressed interested in, supported, and benefited from his presentation on
"what we are doing [in my country] in the food control area." He elaborated:

I presented a seminar on . . . a new concept called hazard and .
. . critical control point. It's a new concept in food processing
where you really have to know the hazards that are associated
with the production of food and the critical points where you
have to control so you can get good products. In the chain from
the raw material to the finished products, . . . you [shouldn't]
need to wait until you've tasted the final product to see whether
it's good or not. But you really have to do the control on the
line. (Post Graduate B)

After his seminar, this student expected an element of exchange, "You tell
them what you are doing, they tell you what they are doing, and then we
can learn from each other." And although the professor acknowledged that
they do not practice hazard and critical control point, "He didn't tell me
what they are doing" (Post Graduate B).

But learning what China is doing is specifically what this student
wanted. He "thought it would be interesting to know what they are doing,
so I can, maybe we can, learn from them or vice versa, . . . but that hasn't
happened." As a result, he "can't even get into the business of comparison
between what they are doing and what we are doing" (Post Graduate B).

Access to Technology

Students linked this lack of mutual exchange to a lack of access. As reported in Section Two, many students were motivated to go to China "purposely to learn some of these basic technologies what we call appropriate technologies from the grass roots." Another student explained that he did not come to China "to learn any complex technology because I can learn that one elsewhere, in the developed world. I've come to learn some of the basic ones, so-called appropriate technology." To this student, appropriate technologies blended traditional and sophisticated methods into cost effective, simple procedures. And although appropriate technologies may be "based on specific principles, . . . you don't have to be an engineer or a scientist to apply [them]." He explained the importance of appropriate technologies:

> These are the concepts we are trying to propagate in our country because . . . most of the farmers cannot afford some of these sophisticated technologies in terms of food storage and food processing. So it's just a step away from the traditional way of doing things to the modern stage of using . . . new advancements in science and technology. [We] are trying to blend the two, the traditional and the more sophisticated technology, and make it cost effective [and] very simple. (Post Graduate A)

Although this student had a very clear definition of appropriate technology, two years into his program, he had not been exposed to that technology. He offered an example:

> [China has] a very simple way of . . . processing rice. I know it's happening because I see them doing it. . . . Such a technology can apply in my country easily, . . . but I have not been given the chance to learn how to do it so that I can go back and teach my people what I'm talking about. (Post Graduate A)

He stated that he is "not very comfortable with this because it would be a complete waste of time to spend almost three and a half years in China and not have the opportunity to have . . . access to that kind of technology . . . [and] acquire the knowledge that I intended to acquire in China" (Post Graduate A).

A second student explained that appropriate technology was also a "really important consideration" for him. As China is "coming up in development right now," he added that "I wanted to see what they are doing . . . and then to see whether or not I could apply that to the situation back at home." He also has not been able to access the technology. As a result, the

student felt that "A large part of the reason why I came to China is not fulfilled. Yes, a large part because this is what I wanted to do before I came
to China" (Post Graduate B).

Access to Sites

In addition to being denied access to technology, students felt they were
denied access to sites. Students "have had no opportunity to visit . . . manufacturing factories and places that
we think will be very good for our course." One student wanted to have
the specific opportunity to see the level of development in the area of food
processing, food technology, and food storage,especially, the storage of
frozen vegetables. He elaborated on the source of these expectations:

> We know [China]. . . exports fresh foods and sometimes semi-
> processed or fully processed food . . . to other countries, such as
> Europe and Japan. And [in] that area [of exports], I work back
> home. So I was thinking that I would have a first hand oppor-
> tunity to see really what they are doing here. Because back at
> home, we usually think that China is a Third World country, a
> developing country, so maybe their . . . the level of technology
> should be more applicable to ours then the West. Yes, so we
> were really interested in seeing . . . their level [of technology]
> and then what actually they are doing in that area. (Post
> Graduate B)

Yet three years later this student was left wanting because he has not "had
the opportunity to go and see what they are doing." He stated, "I just see
the products in the market, but I don't know how they are doing it. There
is no chance of knowing that." The student, however, continued "pressing
hard and hoping that at least they will open up. . . their local industries,
particularly food industries, [for us] to see the kind of machinery . . . and .
. . processing protocol . . . so that we can learn" (Post Graduate B).

For students in Shanghai this issue of access became so critical that they
called upon the General Union of African Students in China (GUASC) for
support. "Under increasing pressure," the GUASC, as "elected representatives of the African student community in Shanghai and the official voice
of the African students," raised the issue of access with the Resident State
Education Commissioner in Shanghai. In an official memorandum to the
Commissioner (Appendix M), the GUASC detailed their concerns:

> . . . we understand that Shanghai is the most industrialised city
> in China. It is really unfortunate that someone can spend 2
> years and [a] half without being given an opportunity to visit
> even a simple factory or manufacturing plant. As mentioned
> earlier we are the leaders of tomorrow in our respective coun-
> tries. Some of us studying here take important eonomic [sic]

decisions on behalf of our Governments in our various jobs at home. We can play a crucial role in initiating vital trade links between our countries and China if we knew exactly what China produces, that would be useful for our countries. Not only are we envoys of our countries here, but would like to give the business community in our countries a clear picture as to what can be bought from China. China is loosing [sic] out in terms of securing a stable market in Africa and indeed [in] other parts of the world by not exposing us to what they are able to offer our markets back home.(General Union of African Students in China, 1996: 4)

By restricting foreign student access to local factories and plants, the GUASC contended that China would ultimately be 'losing out'. However, students also felt they were losing out. Without the opportunity to be exposed to China's industries, they felt "denie[d] in getting the practical aspect of the course." Students reported that classes concentrated on the textbook and while "The theory is useful . . . without the practical aspect of it, [the course] is not complete" (Group of Three, Post Graduate A & B). In the next chapter, levels of academic satisfaction are further explored in relation to language and progress.

Notes

.[1] In regards to the library, students noted the following:
1. "It is hard to get books teachers use in class."
2. "The best books are in foreign languages."
3. "The library does not authorize us to borrow [foreign books]."
4. "I do not have enough time to go to the library because I have to read slowly the academic manuals using   a dictionary" (Notes from four surveys).

Chinese Language and Progress

Chapter Ten investigates issues regarding language. It is divided into three parts: Chinese language training, proficiency, and the relationship between proficiency and academic progress.

PART ONE: CHINESE LANGUAGE TRAINING

Prior Training

Of the 129 students who answered the question about prior language training, only 1.5% (2) indicated that they received *any language training before to coming to China*. One student from Congo noted that he studied the Chinese language "au lycée," and a student from Mauritius "specialized in the Chinese language at home." In fact, this student earned his Master's degree in the Chinese Language and was now studying for his Doctorate in Chinese Foreign Policy. The majority of students, 95.5% (127), however, received no prior Chinese language training. Upon arrival in China, 91.7% (122) received some language training. The length and intensity of the training varied, depending upon the level of the student and in most cases, the language of instruction.

Language of Instruction

Chinese is the language of instruction for all undergraduate programs. All foreign undergraduates undergo at least one year of language training before they start their four-year degree program. If these undergraduates continue into graduate studies, they will continue in the Chinese language. Thus, basically two groups of African students study in the Chinese language: undergraduates and graduate students who already spent at least five years in China doing their first degree.

English is the language of instruction for most graduate programs under China's new 'High Level, Short Period, High Benefit' policy (see Appendix L).

Thus, those who study in English are mainly newly arrived graduate students who did their first degree outside of China. However, some of these new graduate students do study in Chinese.

Chinese Language Training for Graduate Student Studying in English

Some graduate students who study in English receive a few months of Chinese language training while others receive no training at all. There does not seem to be one set pattern. Four Master's students, 3.8%, received no language training. These students, in China from one to over two years, included two Ugandans studying Computer Science in Shanghai, one Sudanese studying Mathematics in Nanjing, and one student studying Plant Nutrition in Hangzhou. While these students had no language training, others indicated that they attended short courses, ranging from two to three months. For example, a Doctoral candidate from Sierra Leone, studying Agro-Environmental Chemistry for the past five years in Hangzhou, had "two months *pinyin*" training. Others studied Chinese for one year but now were taking their courses and preparing their theses in English. Still, others studied Chinese for one year and were now taking their degrees in Chinese. Of those students who did study Chinese for one year, many received their training at the Beijing Language Institute.

Beijing Language Institute

As an *essential element* in these exchanges, one Chinese administrator that I spoke with focused on the Beijing Language Institute and their approach to teaching foreign students. He explained that between 1973 and 1996, the Institute has accepted 3250 students from fifty-one African countries.[1] During this time, the Institute developed many strategies to improve the quality of language instruction. For example, the Institute requires that all instructors speak at least one foreign language, such as English, French, Swahili, Hausa, or Arabic. Thus, teachers, aware of linguistic difficulties related to other languages, can easily locate and track learning problems of students from diverse linguistic backgrounds. Once a problem is located, teachers can use their own foreign language skills to compare and explain grammatical differences. Moreover, teachers understand that students from different educational systems need help to adapt. Thus, teachers strictly ensure that students attend classes six days a week, from 8:30 a.m. to noon. In addition to highly qualified and dedicated instructors, the Institute also provides language learning mechanisms.

Learning Mechanisms: A Chinese Character Inventory

Language experts and researchers at the Beijing Language Institute spent five years creating a Chinese Character Inventory. The Character Inventory divides into three categories: i) Chinese characters in general ii) 2500 frequently used characters[2] iii) 500 fundamental teaching techniques. According to the inventory, students usually require two years of instruction in order to learn the '2500 frequently used characters' necessary for starting their degree. However, after only one year of language training, students might acquire a knowledge of 1500 to 2000 characters. When students enter a degree program with less than the preferred minimum linguistic knowledge, they find the first three months particularly difficult. However, with supplementary classes, students, in time, strengthen their language proficiency (Administrator).

Every Chinese professor and authority I had the opportunity to speak with concurred that language was an *essential element* in the lives of African students:

> [For the students] the biggest problem is language because before they come to China, they don't understand Chinese. The Chinese government requires them to study Chinese for one year for basic knowledge, for daily life language. But for some of them, they cannot learn Chinese quite well [enough] in one year. (Professor)

A colleague at another university also recognized that some students cannot learn enough Chinese in one year. He, however, highlighted that students did improve once in class alongside their Chinese classmates:

> Foreign students . . . have many problems with the language. Chinese students . . . never have this problem with language, so when they study they have a big advantage. . . . But . . . they are in the same class and . . . the same lectures, so the African students improve [their] level, and they can be at the same level as the Chinese students. (Director, Foreign Students' Office)

To "be at the same level" of Chinese students, many universities offer "coaching classes." These coaching classes require extra time and effort from the student, and "the teachers also work hard for it." Teachers also make concessions with exam time. African and Chinese students write the same exams, but if the Chinese students are given two hours, Africans may be given three because of the language (Director, Foreign Students' Office).

Despite coaching classes, exposure to native speakers, and extra efforts by teachers, the prospects of studying for a degree in the Chinese language remains daunting for some students. Often these students go to their embassies for support. All Counsellors discussed the many students that go to the embassy "genuinely [claiming] it is impossible." Students feel that the Chinese language they "learn for one year is not enough for a course."

For example, one Counsellor discussed the case of a student, who, despite her high marks, wanted a ticket home because of the language difficulty:

> I've had a case of one student. . . . Her Chinese language at the Institute was good. She was getting high marks. But this year, she started her course in economics. She is so frustrated. She says, 'I can't follow, so what am I doing here?' Twice she has asked me to get her a ticket to go back. She said the level of Chinese she knows is not enough to follow in economics. (Embassy Counsellor A)

This Counsellor encouraged her, as he did all his students, to persevere and to look at those who came before:

> I've always told her, '. . . Just give it a chance, another chance. All these students who have studied here, who have finished, have had the same difficulties. It's not really a unique case. Your marks were almost the best at the Institute of Languages. . . . So how about those you are beating in Chinese? They are here doing their degrees in Chinese too. So you are not alone. Just give it a chance. It will come.' (Embassy Counsellor A)

A second counsellor also acknowledged the many students who come to him to discuss the "degree of difficulty in [the] Chinese [language], and [report] they cannot follow." He also encouraged them to persevere and to look at those who came before for inspiration:

> I always advise them, 'Be patient. All of these fellows you find here had the same problems, and they are going ahead. Those who finished before you, they had the same problems, so give it time. You'll manage.'

> And gradually, they tend to manage. I think they have. They are not complaining anymore. Maybe they, you know, at times when they talk between each other and [with] other fellows from other countries, they find it is the same problem. So they settle down to the reality that it is going to be like that. (Embassy Counsellor C)

While some students 'settle down,' others would like to change that reality. On one questionnaire, an undergraduate student advocated longer and specialized Chinese language courses: "Though we take language lectures first it is not enough to enable us to study smoothly. . . . The language [training] needs more time, . . . more than one year, or I think the language should be taught according to one's speciality, not generally." He ended his note optimistically, however, adding that despite these difficulties, "People try to come up over time." Although this first year of language training and the prospects of doing a degree in Chinese may be overwhelming, clearly, people do "come up over time." Undergraduate students who spend one year in a language course may begin their degree programs on unsure footing, but by the end of a four-year degree program, these students become proficient, as the following statistical data indicates.

PART TWO: CHINESE LANGUAGE PROFICIENCY

As previously discussed in Chapter Four: Research Design and Data Collection, I initially prepared the questionnaire for three subject populations: Engineering, Medical, and Agricultural students. Thus, when I asked the students to rate their proficiency in the Chinese language and discuss the relationship between their proficiency and their academic progress, I had these very specific populations in mind. These students would have all been studying their degrees in Chinese. As these intended subject populations changed, two characteristics among the new, unanticipated populations emerged for which these questions on Chinese Language and Progress did not fully apply.

First, thirty-seven of the questionnaire respondents turned out to be newly-arrived undergraduate and graduate students. Out of the 133 students, twenty-four had been in China for less than five months, nineteen had been in China for just one year, and four had been in China for one and a half years. These questions about language proficiency did not fully apply to these students who had just arrived or had just finished their language training and barely begun their degree studies. Some students just left the section blank while others wrote a note to explain, "Personally, I think it is too early to judge." Second, many questionnaire respondents turned out to be graduate students studying in English. Thus, for these students, this section did not fully apply as well.

Adjusting the Data

In light of the two new elements in the subject population (new students and students studying in English), I adjusted the statistical data file to make the findings in Tables 10.1 and 10.2 meaningful. Thus, for Section 6 of the questionnaire, questions 3-5 only, I removed all statistical data from students who had been in China for less than one and a half years, and I removed all students who studied in English.

Language Proficiency of Students Studying in Chinese

The following findings were thus taken from the remaining seventy-two students who had been in China from two to thirteen years (mean of 5.035 and median 4.00 years) and were studying for their degrees in the Chinese language. Of these seventy-two students, 37.5% (27) were studying for their Bachelor's, 43.1% (31) for their Master's, 13.9% (10) for their Doctorate, and 4.2% (3) were studying for a diploma or specialized certificate. In this sample, all graduate students did their undergraduate degrees in China. As presented in Table 10.1, these seventy-two students rated their ability in the Chinese language on a scale of 1 to 5, (Fluent,

Advanced, Intermediate, Fair, Poor), according to four skills areas: reading, writing, listening, and speaking.

TABLE 10.1
Self-Rated Language Proficiency
of Seventy-two Students studying for Degrees in the Chinese Language
by Frequency (FQ) and Percentage (%)

	Fluent		Advanced		Inter-mediate		Fair		Poor		No Response		Totals	
	FQ	%	FQ	%	FQ	%	FQ	%	FQ	%	FQ	%	FQ	%
Speaking	28	38.9	18	25.0	17	23.6	9	12.5	0	0	0	0	72	100
Reading	11	15.3	23	31.9	22	30.6	14	19.4	2	2.8	0	0	72	100
Writing	7	9.7	14	19.4	23	31.9	18	25.0	10	13.9	0	0	72	100
Listening	20	27.8	23	31.9	17	23.6	10	13.9	1	1.4	1	1.4	72	100
Overall Ability	11	15.3	21	29.2	24	33.3	11	15.3	0	0	5	6.9	72	100

Of all the four skill areas, students rated themselves most highly in speaking and listening. The highest rated skill was *speaking*: 63.9% (46) considered themselves 'fluent' and 'advanced'. The second highest rated skill was *listening*: 59.7% (43) considered themselves 'fluent' and 'advanced'. Reading and writing seemed to be more difficult areas. While 47.2% (34) considered themselves 'fluent' and 'advanced' readers, only 29.1% (21) considered themselves 'fluent' and 'advanced' writers. One student, however, noted that his writing was "improving by using the computer." *Overall,* 15.3% (11) of students considered themselves 'fluent', 29.2% (21) 'advanced', 33.3% (24) 'intermediate', and 15.3% (11) rated their overall language abilities as 'fair'. No student rated their *overall* language ability as 'poor'.

These statistical findings correspond to observations made by a Director of an International Exchange Division. In an interview, he praised African students' linguist capabilities, stating that out of all of the foreign student, Asian students might have an advantage in reading and writing, but he found that the speaking and listening fluency to be the highest among African students.[3] The Director attributed the Africans' fluency to multilingual exposure and psychology. He pointed out that in one African country, such as Zambia, it is possible to find more than seventy-three different languages. Thus, he remarked, that many of these students grow up in a highly multilingual environment. In fact, data from Section One of the questionnaire revealed that the population was remarkably multilingual. Over 88% (115) of these students were at least trilingual.[4] In addition to this multilingualism, students' "psychology," according to the Director, was also a contributing factor. The Director found students were generally "not timid [and] not afraid to make mistakes." In his thirty years of experience, he found that among all foreign students, the Africans were the most fluent speakers of the Chinese language (Director, International Exchange Division).[5]

PART THREE: ACADEMIC PROGRESS AND LANGUAGE PROFICIENCY

Of the seventy-two students studying for their degrees in the Chinese language, the majority, 51.4% (37), reported a 'moderate' degree of *satisfaction with academic progress* while 30.6% (22) felt 'great' *satisfaction*. At the same time, 36.1% (26) felt that their *linguistic proficiency in the Chinese language [was] related to their academic progress* to a 'moderate' extent while 34.7% (25) felt that language and progress were related to a 'great' extent. Table 10.2 highlights this data.

TABLE 10.2

Relationship between Academic Progress and Proficiency in the Chinese Language of Seventy-two Students Studying for Degrees in Chinese by Frequency (FQ) and Percentage (%)

	Very Great		Great		Moderate		Little		Not at all		Totals	
	FQ	%	FQ	%	FQ	%	FQ	%	FQ	%	FQ	%
*1. Satisfaction	4	5.6	22	30.6	37	51.4	7	9.7	2	2.8	72	100
*2. Relationship	13	18.1	25	34.7	26	36.1	7	9.7	1	1.4	72	100

1. Satisfaction: Extent of satisfaction with academic progress
2. Relationship: Extent of relationship between Chinese language proficiency and academic progress

Chinese Proficiency & Academic Progress among Graduate Students with English Option

By contrast, most students studying in English found their Chinese proficiency 'not at all' related to their academic progress. Many students, who had been in China for a few years, rated all of their language skills as 'poor' but found this poorness had no bearing on their academic progress because their courses were in English. This type of case came up again and again. For example, the previously mentioned Doctoral student from Sierra Leone who had two months *pinyin* training reported that after five years in

China, his speaking and listening skills were 'intermediate,' but all other skills were 'poor'. Nevertheless, he felt 'very great' satisfaction with his *academic progress* and considered his language skills 'not at all related' to his studies. A Master's student in China for four and a half years, indicated that he had one year of Chinese language training, but his degree program was in English. Thus, while he rated his *reading* and *writing* as 'poor' and his *speaking* and *listening* as 'fair', he considered his proficiency to be 'not at all related' to his 'great' *academic progress*. On the bottom of the page, this student wrote, "Generally, when you do not study in Chinese then the Chinese language is only useful here in China." This case and types of comments were common. A Master's student from Ethiopia indicated that he received one year of language training "because I was forced." He emphasized this statement by an asterisk and then wrote it again at the bottom of the page. This particular student rated all of his language skills as 'poor' and his academic progress as 'poor,' but at the same time he felt that this poor progress was 'not at all related' to his poor Chinese language skills because his program was in English.

Although these students studying in English found no relationship between their academic progress and proficiency in Chinese, on the questionnaire and in the interviews this particular group raised many issues in relation to language. Students' comments were plentiful and consistent, and while definite trends and patterns emerged, no clear, uniform picture developed.

Language Expectations

On one questionnaire and in two separate interviews, this issue of "forced" Chinese language lessons was brought up by students from Ethiopia, Ghana, Namibia, and Sierra Leone. Students expressed concern that they agreed to go to China with the understanding that their programs would be a certain number of years and in English. They were taken aback when, upon arrival in China, they learned that they would have to spend an extra year studying Chinese. One Ghanaian expressed concern by "the fact that I had to spend one year in Beijing to do [a] one year Chinese language course which was not in our original plans before coming here." With family and employment responsibilities waiting for him at home, he felt that this extra year of time was a serious imposition. Moreover, he, like all the others, felt troubled that they were "forced" to spend this unexpected, extra year to study Chinese.

At the same time, students were further concerned because they felt they could not possibly acquire sufficient language skills to properly pursue graduate studies within this one year of training. One student spoke for many when he stated:

> I came to see it was just impossible for somebody to come here and do
> one year Chinese and go to do his Master's or Doctorate, using Chinese

characters and writing a thesis and everything in Chinese.

. . .Yes, to me, I still insist, I've done the course [in] one year, I did well [the] exams, but that doesn't make me prepared enough to do my Master's in Chinese. (Post Graduate B)

Fundamental Questions

According to every Counsellor I spoke with, this situation and reaction were not uncommon. Time and time again, students went to the embassies to discuss the feasibility of studying Chinese for one year and then pursuing a university degree. Moreover, for the Counsellors and many students, these language issues posed an even more fundamental question: namely, the quality of the degree. One Counsellor described a "mature" student who came to the embassy "wondering [and] asking what kind of degrees, . . . what quality of degree, . . . [was he] likely to get, . . . especially, if he is doing it in a language that [he doesn't] seem to understand?" This student asked the Counsellor:

> If they give me a test now, what do I write? So if I spend a whole year like this, that's almost a year wasted. At the end of the day, am I really going to do research in this language? Am I going really to write a good thesis in this language? Am I really going to acquire enough knowledge in itself ? . . . If I cannot consult textbooks, if I cannot do [research], how do I know the Chinese are not giving me free marks? What [is] a really good degree here? (Embassy Counsellor A)

These fundamental questions about the feasibility of studying Chinese for one year and the "quality of the degree" pursued thereafter were ones that the Counsellor himself had "always wondered" about, especially "if [students] don't have enough language to do research in their books . . . and read really widely." He admitted that he has "always questioned the quality of the degree that they get." (Embassy Counsellor A).

While this Counsellor raised serious questions about the feasibility of new graduate students doing academic work in the Chinese language, he also questioned the abilities of students who had been in China for many years. He acknowledged that while they do speak fluently, he had his doubts about their abilities to read and write:

> A student may finish here a course in Engineering, but if you give him a Chinese simple letter to read for you, he will not manage. Because they speak it very fluently, they speak it very well, but [if] you find an announcement for an art exhibition somewhere and you ask him, 'What it is?' . . . You will find, they won't tell you. (Embassy Counsellor A)

To illustrate his point, the Counsellor offered an example of a student who reported to have earned his degree orally:

A student from another country, a friend really, quite close to me, . . . he finished his degree in fine arts . . . [and] speaks Chinese very fluently, married to a Chinese girl. . . . But when he passes a shop somewhere and you say, 'What is written there?', he cannot tell you.

[I say], 'You are supposed to have done your degree in Chinese. How did you do it? Is it a free one? Did they give you a free degree or what?' So he says, 'No, ours is more practical'. But I say, 'No. In art there are written papers everywhere, the history of art, all that kind of stuff. How do you write it?' He says, 'No, no, I prefer to do it orally.'

He convinced me that he did it [orally], but then that means, 'How about your course work, was it always oral?' And so I found there was some flexibility somewhere. (Embassy Counsellor A)

Because of this "flexibility" and concerns around the length, feasibility, and quality of the degree, many students lobby to have their programs in English. In fact, according to another Counsellor, the issue of finding schools where English is the language of instruction was one of the main concerns students raised in his office. Students came to the embassy to discuss "the language they face" and to ask, 'Is it possible we do it in English? Is it possible?'" (Embassy Counsellor C). The Counsellor admitted that "Some of them get scared and take off," though he encouraged them to stick to Chinese. He stated, "Most of the time, we tell them, 'Look, take the [course] . . . try and learn, . . . probably the Chinese language will be useful tomorrow. Don't look for English only. . . . Try to have another language'" (Embassy Counsellor C).

Despite such encouragement, however, many students do look for English programs. Students feel that if they were taught in English with English materials, "They should be able to follow better, and . . . at the end of their degree, feel that really they worked, and they are getting a good quality degree" (Embassy Counsellor A). In fact, the same Counsellor described a case where a number of students came to the embassy and announced:

We are already forming a group to go and complain to the Chinese [authorities]. We can't follow. We cannot. It is impossible. They have to teach us in English. . . . [We] are insisting. They have to teach us in English. We can't follow in Chinese. (Embassy Counsellor A)

Schools and programs that offer an English option become attractive for many. One student, who spent an unexpected year in Beijing, "insisted . . . on a school that would take my course in English." He was not alone:

Actually, a lot of people here complain, a lot of people have changed universities and courses just to have the chance to do their courses in English. So consider . . . for instance, we have a lot of foreign students

... in this university, a lot, because of the fact that most of their courses are taught in English. (Post Graduate B)

Catch 22

Many students do protest that Chinese instruction was not in their original plans and successfully lobby to study in English. However, they then encounter what this same student described as a "Catch 22":

> You know, after going ahead and doing my course in English, you go and meet professors, lecturers, and teachers who know very little English, at least in speaking. So you find out that you cannot communicate well with your professor, your lecturer, during class hours because his English is not so good, and so he can't express his views or explain things well.

> You see, so it's a bit frustrating actually because you know he wants to do his best, you know he wants to tell you what he wants to, but he can't do it because of the language . . . difficulty. So that's another point. At least as of now, it's still a vital, bitter point many students are facing. (Post Graduate B)

On the questionnaires many students raised this "vital, bitter point." One student declared, "Language is a huge problem here. Those of us taking our course in English often meet lecturers who know little or no English. Those using the Chinese language have even a bigger problem because of difficulties in grasping well the language." For a Kenyan student, this problem distracted from the benefits of the program: "The courses are well-designed, the laboratories are well-equipped, but communication in English is a major bottleneck. One, therefore, should put more efforts to bridge the gap."

Language was a difficult issue for students, but students also recognized the difficulty for professors. A Ghanaian captured the sentiments of many when he wrote: "Teachers [who] teach in English should be given orientation or proficiency training in the language and should not use students to polish their English." While students were critical, they were also empathic with professors "who want to do their best." One student described what happens in the classroom when the communication breaks down:

> [Professors] keep on apologizing, and they keep on apologizing about it, that, you know, they just can't speak. They just can't explain. Well, sometimes they just give up because they can't do it, and they also get frustrated actually. So you find yourself having to do most of the work yourself. (Post Graduate B)

Thus, he described the dilemma: "Do it in English, . . . it will be very difficult for the lecturers, and do it in Chinese, and ah . . . I wouldn't even think of it. So it's like a Catch 22"(Post Graduate B).

Final Thoughts on Language and Cultural Cooperation

In closing, one Counsellor reflected upon the issue of language proficiency and feasibility in relation to cultural cooperation and in a way brought up another type of Catch 22. The Counsellor pointed out that in order for the cultural cooperation to be successful, students must be successful:

> If it is a cultural cooperation, those students who come here must finish. . . . If they [all] have the same level of difficulty, you must find a way of standardizing the courses and the tests, so that at the end of the day, they all go home with the qualification. Otherwise, you will be explaining [to] them every other day, and there will be no use of coming here now. There will be no use for this exchange.

> This is what I want to mean. My feeling is that there is this kind of flexibility. You are a foreigner. You have this level of difficulty [with the] language, so the exam must be oriented towards this difficulty. So I shouldn't require much from you because is most cases you don't even understand what I teach you. So that at the end of the day, a serious student . . . may feel that he may not get the kind of knowledge he came to look for. (Embassy Counsellor A)

These Catch 22's are not limited to issues of language. In fact, the next chapter on Financial Support reveals another paradoxical situation where students on a "full scholarship" struggle to meet a fraction of their needs.

Notes

.[1] The administrator pointed out that the figure 3250 does not represent the total number of Africans who studied in China because African students were sent to many other universities and institutes across the country.

.[2] As a point of reference, the administrator explained that a knowledge of 3500 characters enables people to read and understand 97% of the newspaper.

.[3] Snow (1988: 205) emphasizes that such praise is "not entirely flattery. Several African languages such as Hausa and Kikuyu are spoken, like Chinese, in a range of different tones, and partly for this reason African students often learnt to speak Chinese faster and better than their Western counterparts."

.[4] For this conclusion I considered that in answer to questions 4 and 5 of Section One: Profile, the majority of respondents, 75.2 % (100), identified one mother tongue, 19.5% (26) identified two, and 3% (4) identified three mother tongues (3 out of 133 did not respond). Moreover, 12% (16) of students also identified one other language that they spoke in addition to their mother tongue, 30.1% (40) identified two others, 32.3 % (43) three others, 17.3% (23) four others, 5.3% (7) five others, and 1.5% (2) identified six other languages that they spoke. Thus, to reach the conclusion that 88% of these students were at least trilingual, I added up the number of students who spoke two or more languages in addition to their mother tongue (40 + 43 + 23 + 7 + 2 = 115), and then I divided this figure by the total number of respondents to this question (115 -131= 87.7%). It is also useful to note that for most of these students their linguistic knowledge is derived from at least three broad geographic areas: Africa, Europe, and Asia.

.[5] It is interesting to think about the linguistic accomplishments and potential of these students in relation to Ali A. Mazrui's *World culture and the search for human consensus* (1975b) and his "structure of peace" found therein. Mazrui's structure of peace is partly based on five languages: English, French, Russian, Arabic, and Chinese. Mazrui defines Chinese as an "irresistible" world language in that it is spoken by "a fifth of mankind" and links the elevation of the Chinese language to cultural and political possibilities for new social directions (1975b: 23-37).

At that time, Mazrui felt that Chinese had "certainly not crossed continents except with overseas Chinese and some Western scholars."(1975b: 24). Now almost twenty-five years later, Mazrui might find some promise in the linguistic accomplishments of these African scholars.

CHAPTER ELEVEN

Financial Support

This chapter investigates the sources, approximate percentages, and sufficiency of students' financial support. In Section Seven of the questionnaire, students were first asked to indicate all sources of funding and approximate the percentage from each source to account for 100 percent of their income. If students felt their income was insufficient, they were asked to explain.

PART ONE: SOURCES AND PERCENTAGES OF SUPPORT

As Table 11.1 reveals, the largest majority of students, 87.2% (116), received an average of 51.7% of their income from the *Chinese Government*. Students indicated that those contributions ranged from 3-100% of their total funds. In addition, a majority of students, 63.9% (85), indicated they received an average of 47.12% of their financial income from the *Government of their own country*. Those contributions ranged from 5-100%.

When examining Table 11.1, two points should be kept in mind. First, a few students noted that the percentage of financial income that they indicated represents the amount received in hand. In other words, the figure, as one woman specified, "doesn't take into account the amount dedicated to the other domains of academic life (which I don't know the value)." The "other domains of academic life" include airfare, tuition fees, and lodging in a semi-furnished room. These costs are covered by the Chinese scholarship. My feeling is that when answering this question, most students indicated the amount of money they receive in hand because the values of these other domains were unknown.

Second, as this chapter will detail, many students felt their support was so "wholly inadequate" that when they answered this question, they approximated percentages that purposely did not equal 100 percent to emphasize the point that they felt severely underfunded. For example, one student indicated that she received 5% of her income from the Chinese Government and 15% of her income from the Government of her country, and she carefully totalled this amount to 20%. In other words, she could not see fit to indicate that what she received could possibly equal 100% when it felt more like 20% of an income. For these reasons, the figures in Table 11.1 do not add up because students felt their sources of income did not add up.

TABLE 11.1
Sources and Amounts of Financial Support
by Frequency (FQ) and Percentage (%)

	1. Students with this source of income		2. Average contribution from this source (in%)	3. Range of contribution from this source (in%)	4. Median contribution from this source (in%)
SOURCES OF INCOME	FQ	%			
Chinese Government	116	87.2	51.67	3-100	50.00
Own Government	85	63.9	47.12	5-100	50.00
Family	35	26.3	26.14	5-100	20.00
Personal	21	15.8	20.47	5-70	10.00
Agency	3	2.3	28.33	5, 10, 70	10.00
Other	2	1.5	30.00	10, 50	30.00

KEY:
1. **Students with this source of income:** Students checked one or more 'sources of income' as it applied to their situation.

2. **Average contribution from this source (in %):** Students approximated the percentage of income received from each source.

3. **Range of contribution from this source (in %):** Some students, for example, approximated the percentage of income they received from the Chinese Government was 3% while others approximated 100%. Thus, the range of contribution from the Chinese Government was 3-100%.

4. **Median contribution from this source (in %):** This refers to the middle value (with values of equal total below and above) of the contribution from each financial source.

PART TWO: SUFFICIENCY OF FINANCIAL SUPPORT

The majority of students, 71.4% (95), reported that their financial support was insufficient while 24.1% (32) indicated that their resources were sufficient. Most of those 24.1% with sufficient funds were graduate students who study as part of their employment training. In the questionnaire and in the interviews, these students acknowledged their privilege:

> The Chinese scholarship is woefully inadequate. I, we are able to survive here because of what we receive from our government and our employer but not because of what we receive from the Chinese [Government]. (Post Graduate A)

A Zambian man revealed that he also "survives" because of a similar arrangement:

> I work for my Government back home and am on paid study leave. The [stipend] from the Chinese isn't sufficient. Very often I have to call for my salary back home for 'buck up.' (Survey Note)

While these students managed because of their salary from back home, the majority of students found their financial resources to be "wholly inadequate." Out of 133 questionnaires, more than half (68), wrote comments about their financial situation. Only one person wrote that he found his resources to be "satisfactory without luxury." One other indicated that he felt the stipend was insufficient, but this insufficiency was not all bad:

> The inadequacy of the stipend is, in my honest opinion, good for most African students. . . . Given the terrible managers of ours resources that we are back home, some financial discipline could arise from this situation and may be pilfered to other sectors of our everyday life when we get back home and develop our nations to stir away from the perpetual dependency syndrome. (Survey Note)

All other written comments claimed that the insufficiency of financial resources was a serious problem. To ascertain the claim, the General Union of African Students in China (GUASC), in the previously mentioned memorandum addressed to the Resident State Education Commissioner on July 4, 1996 (Appendix M), presented a survey of students' monthly requirements. The survey detailed the costs of meals, stationery, and toiletries for one month and found that the "the barest minimum that can meet the barest minimum of daily need" was 1185.00 *rmb*.[1] The GUASC "pleaded with the Chinese Government to look at our plight with sympathy if we have to survive" and proposed an allowance of 1050.00 *rmb* per month. At the time of this study, undergraduates received a monthly stipend of 500 *rmb*, and graduate students received 650 *rmb*. Students expressed the extent of their discontent in the questionnaires and interviews.

The Full Scholarship: Financial Support from the Government of China

First, the issue most often raised was that students went to China understanding that their scholarships were full, and they would be well provided for. Students reported this was not the case. Students wrote in detail about the "so called" Chinese scholarship. One student noted, "When they call us from our countries, they pretend they give full scholarship[s]." Another understood that, "China promised to take us in charge, but in reality it does not do that." On the understanding of this promise, another student wrote, "I sacrificed my good job at home for nothing." Two others noted that "The Chinese authorities said our scholarship [was] full, but what we are getting [does] not commensurate to what is expected by a foreigner to live an average life in China." In short, one concluded that students "[are] not getting anything that was spelled out in our letter, . . . so they are deceiving us"[2] (Notes from eight surveys).

All Embassy Counsellors I spoke with were well aware of these concerns:

> The Chinese tend . . . to tell us, as a Government, that the scholarship is complete; it is a full scholarship. Whereas actually, on arrival in this country, okay, you find that the accommodation and tuition are covered, but the extra money, the living allowance is really inadequate, very inadequate, very, very inadequate. (Embassy Counsellor A)

Students expressed a sense of feeling misled and wrote that authorities should "make students fully aware of what scholarship entails." In fact, one Kenyan student declared that "the biggest problem is in getting the correct information [about] the 'full' scholarship." In an interview, students discussed this problem and the confusion around the definition of a full scholarship:

> If you applied to Western universities, you know your tuition fees, you know [the] allowance fees, you know your everything fees. Here, no way! Nothing is clear.

> A [foreign] student in China will not tell you how much tuition fees [are], how much books cost, . . . and in this respect it is very hard for a student, an African student, to know what does full scholarship [mean]. Can you give me the definition of a full scholarship? Can we say it is a full scholarship when someone cannot join the two ends, the two extremities of the month?

They pretend . . . it to be a full scholarship. Can I call it [a] full scholarship when I can't reach the 10th of the month? (The Four Graduates)

A student from Ghana agreed when he wrote, "Blanket statements [such] as full scholarship . . . lack understanding and should be rectified if the cooperation is to stay." The same student recommended that this problem may be solved if the "Chinese side . . . spell out in clear terms the content of the scholarship, stating what they can offer and what they cannot" (Survey Note).

UNESCO

One result of this confusion is a highly charged debate about the involvement of UNESCO. On one questionnaire and in three interviews, students brought up their understanding that these scholarships from China are in fact from UNESCO. According to the students, "The money . . . is not from China, . . . they are managing [it] from UNESCO":

> The Chinese government is developing, kind of like us. They are poor, like us. . . . It's only UNESCO, through [the] Chinese government, [that] gives me this scholarship. The Chinese government is only managing. . . . UNESCO [is] how we get our scholarships. (The Four Graduates)

In other words, students claimed that UNESCO pays China to educate them. Declaring, "This UNESCO thing has to be cleared up," another student explained:

> Our countries are . . . members of the United Nations. The United Nations has many branches, including a UNESCO branch for United Nations for Science and Education and Technology. This UNESCO is a program [among] one of the UN programs. UNESCO [looks] among those members [of] the United Nations [which are] developing countries [and] helps them get scholarship[s] in order to educate their future leaders.

> UNESCO invested [in] education, and they wanted to put [the investment] in China just because China was claiming that it had infrastructures, universities, laboratories . . . to host, to receive African students here. [UNESCO] sent African student[s] to China [because] China['s] . . . living costs were relatively cheap [and thus they could educate] the maximum [number of] students.

> When China became a member of the UN, they wanted to show they are supporting the cooperation South and South. And they wanted to show, they are helping [a] special, old friend, the African countries because [then the UN] helps the Chinese government to excel in their transitions. They say, 'We are *helping* African countries.' [But]

these things are not help, [it's] just managing. (The Four
Graduates)

In addition to the claim that China was "just managing" the programs, students expressed concern that the Chinese Government and in some cases
their own Governments received money from UNESCO on behalf of students and failed to give it to the students. In other word, students charged
that the Governments profited at their expense:

> Under UNESCO, contracts have been made between [African] govern-
> ments and China. We don't know how they share commissions, [but
> African] government[s] and [the] Chinese government get [a] percent
> [from UNESCO]. (The Four Graduates)

However, as convinced as these students were about the role of UNESCO
and the Governments' profiteering, the three Counsellors I spoke with were
equally convinced otherwise. Aware of these beliefs among students, one
Counsellor commented:

> People believe that [UNESCO] gives [money] to the Chinese who are
> taking the money. . . . Some students say UNESCO is giving supple-
> ments to countries, so that the money should come to them and that
> the money is kept. It is wrong. I think it's wrong. It is a bilateral issue.
> It's between China and the [individual] countries. . . . It's really a
> rumour. (Embassy Counsellor B)

Another Counsellor also dismissed these claims as rumour:

> UNESCO has no hand in this. It's highly bilateral. . . . These scholar-
> ships existed even when China was still enclosed. UNESCO didn't even
> have an office yet. [So] no, I don't think so. I think it is Chinese money.
> I don't even think UNESCO is aware of that. I am very sure of that.
> (Embassy Counsellor A)

The third Counsellor concurred with his colleagues and claimed he had
"never heard of [UNESCO's involvement]. . . . For [our] students, China
totally sponsors all of them" (Embassy Counsellor C).[3]

While there was no consensus between the students and Counsellors on
UNESCO's role, the entire scenario, be it a rumour or be it true, indicated
a level of confusion. For the students, this confusion was further exacerbated by the lack of support from their own countries.

Supplementary Funds: Financial Support from Home Governments

Students expressed confusion over UNESCO, felt misled by the promise of
a "full scholarship" from the Chinese Government, and at the same time,
felt disappointed that their own Governments did not support them on a
consistent basis. A few countries, not all, did promise supplementary funds.

This helped. One Burundian wrote that the Chinese scholarship was insufficient, "but with the help of my country, it is bearable." Yet more students, like this Cameroonian felt, "We are never sure that the money will come from our home country." His country mate added that "The funding from my own country comes with a big delay or simply doesn't come at all." A student from Congo similarly found that "Bursaries [from home] come late and sometimes don't come at all." A Zambian also echoed this concern: "The stipend I get from my country comes irregularity and cannot be relied upon."A Sierra Leonean attributed this unreliability to national difficulties: "Support from home is not forthcoming due to the present political instability." A Malian student had "not received any bursary supplement from [his] country of origin . . . since September 1996" while another had not received any money at all stating that, "The Government of my home country has never paid its part of the bursary for all the students in China." A student from Benin wrote, "If I agreed to come, it's because my country had guaranteed that supplement. But ever since I arrived in China, I have never received it." He added that all students "who do not have a bursary supplement from [their] own country are living a very miserable life in China" (Notes from nine surveys).

While the home Government of these students may have had some intention of providing extra funds, most students never expected any money from their governments. One student from Mali wrote that "Since the Chinese government announces that the bursary it gives us is a full scholarship, so our country does not give us any expense money." Another student added that because his scholarship was assumed to be full, he had "no reason to seek alternative funds." Finally, one student ended with a personal plea to home governments:

> I would like to appeal to the African governments [regarding] the cost of living here in China. . . . Give sufficient financial means to African students who come here because life is very expensive here for the moment, and that is the prerequisite for studying well. (Survey Note)

One Counsellor explained the various circumstances students may find themselves in relation to their home country. Some countries promise 1000 *rmb*, and students get it. Others countries promise 3000 *rmb*, and students never get it. Other countries promise support on the condition that the country is doing well. Other countries simply never promise. This Counsellor felt that students who come without false hopes and expectations were the best off (Embassy Counsellor B). Another Counsellor knew of "only a few countries that manage to keep their students a little bit more comfortable than the others. . . . [In fact,] most African countries now cannot afford the extra funds." He admitted it was "a common problem we share with the other diplomats."

Diplomats face this problem directly as many students "approach the embassies asking not [for] individual help but [for] intervention from the government back home." Although the embassy agrees with students and feels it is "a genuine question," the embassy "cannot help them." Instead the embassy offers moral support, "You just have to sit down and study with them the problem. Tell them the background [of] where they are coming from. And they get contented, but really, they are really miserable, that's for sure" (Embassy Counsellor A).

Inflation

As many students pointed out, this misery comes at a time of "exceptional growth [in] China." As another student put it, China is "developing every day." Many wrote about China's "economic evolution" and ensuing "skyrocketing inflation." One student noted that "Prices change every year, but the resources don't change." Another specified these claims: "Prices for all products and services have gone up about 400% [and the] level of life has changed, [but] the bursary remains the same." (Notes from seven surveys).

The General Union of African Students in China (GUASC) asserted that the vulnerability of African students in China is one side effect of the economic growth:

> . . . China's economic reform implementation is now the world's fastest. Any fast economic growth is never without its side effects. Among the vulnerable have been students on scholarship. Adjustments in prices and general cost of living to march [to] economic reform demands, entail new hardships on the part of the student. This now means that the students can hardly manage to afford the basic necessities of life. . . . (General Union of African Students in China, 1996: 2)

Making Ends Meet

With inadequate scholarships, little support from home, and economic inflation of 400%, students reported that they were unable to afford the "basic necessities" of life. They expressed further frustration in that, unlike local people and other foreigners, African students have "no access to part time jobs." They described the "impossibility of carrying out lucrative work . . . even during holidays to make up for the low allowance received as stipend." Again and again, students lamented that for them there was "no possibility of earning extra income." (Notes from five surveys).

Students characterized their financial situation as "desperate." Many wrote about the problems of sustaining themselves with money "just enough" or "not even enough for food to eat." Some students wrote that the scholarship was "just enough for surviving" while others found it "difficult to survive on." Some wrote of the "impossibility . . . [to] join the two extremities of the month" while some conceded it "just allows them to join the two ends of a month." Counsellors recognized and acknowledged these concerns as "genuine":

I've been to their canteens. . . . I know the cost of their food. If [they] are spending very less than reasonably, [they] may just be spending about 20 yuan per day. So that is . . . 600 [yuan]. For undergraduates, it's already gone beyond. (Embassy Counsellor A)

With the money "gone beyond," students felt they were not "permit[ted] to eat well, [let alone] think of buying books, clothes, etc." Many added that clothes, train tickets, recreational activities were out of the question. Two students also noted that they could not even afford "medical services since students are supposed to buy them first and get reimbursements later." Moreover, students claimed that they could not effectively pursue their academic endeavours because stationery, class materials, books, and research supplies were simply too expensive (Notes from nineteen surveys). This particular point was taken up by the GUASC:

> . . . the issue of book allowance and research fund[s] needs attention. Some universities do not offer any provisions for book allowance, while some give a bare 200rb [rmb] per year. This is inadequate. The problem of book allowance needs attention. This is especially critical among students studying in English. There are very few relevant books written in English, and when such books are available, they are very expensive. We are committed to our study, and we feel that we can only derive maximum benefit from it if we have enough learning materials. Apart from the issue of book allowance, the other problem is faced mostly by graduate students, who have to carry out research projects before they can graduate. To carry out sufficient research work in some cases requires extensive travel and purchases of materials. There is a severe disparity in the funding given by various universities for research work. While realising the varied nature of the various research programs, we feel that funding in some cases requires review, because it falls far below the minimal requirements for carrying out any meaningful research project. (General Union of African Students in China, 1996: 5)

While the GUASC made the point that under funding inhibits students from effectively pursing academic endeavours, other students reported that under funding also inhibits them psychologically. One student wrote that he cannot "effectively pursue my academic plans and program because I am worried of how to survive for the rest of the month when the money is finished." Another felt that poverty forced students to "stay in our ghetto and as a consequence [this] deprives us of good knowledge of the outside milieu, outside the campus." In addition, three other students wrote about feeling stuck and cut off. One expressed regret about being unable to afford "any communications (telephone or fax)" while another specified: "I have not even once called my parents, even I cannot frequently write a letter to them because I cannot afford [it]." A Tanzanian woman felt that her inability to afford such communication services added to her melancholy:

As a foreign student in China, I feel the need to be in contact with the
outside world every now and then — to know what's going on out
there (back home especially). But most often these facilities (phone,
fax, e-mail) are often VERY EXPENSIVE for a normal foreign student
to afford. So I guess it adds to the loneliness, depression, and home-
sickness. (Survey Note)

Discrepancies in Prices

Students expressed further distress because "foreign students are not
tourists here, but in China students are supposed to live like tourists, ie.
accommodation in designated hotels, soft seater in trains, and expensive air
tickets, not to mention expensive visa renewals . . . just like tourists." In
addition to tourist costs, students pointed out the "differences made
between foreigners and Chinese citizens [in terms of the] prices of goods."
Another explained that "Foreigners are to pay for the same services more
than what Chinese pay." Students wrote about feeling "swindled," especially when shopping because "[merchants] multiply the prices by two"
(Notes from four surveys).

Discrepancies in Funding

Moreover, on three surveys and in two interviews, students raised other
discrepancies when they spoke of "realizing" that there was "a certain
injustice towards the Africans, if we compare what China does for the
Asians, Europeans or Americans." A woman specified this realization
when she wrote that "The amount of allowance given to African students
[is] less than what [is] given to others." Another student detailed this claim:

> All foreign students are not getting the same allowance. They give
> African and Arab students less money compared to the same students
> under the same scholarships from Europe, America, Korea, and Japan.
> These students from these countries are given more money, more facil-
> ities, and [are] more highly respected than the Africans.4 (Survey note)

Respect

Respect was an issue raised when one student pointed out that the questionnaire did not ask "if China responds well to the total sponsorship of
the foreign students." It seems the response was not positive. According to
one, "The respect to scholarship students is getting less . . . . Self-finance[d]
students are welcome[d] at high[er] respect." Respect was also on the mind
of another who wrote, "According to the Chinese, Africans . . . come to
beg here in China . . . only because they do not have enough food in their
own countries, which are very poor." (Notes from three surveys).

While the GUASC contended that financial vulnerability of African students was an economic side effect, other students asserted social vulnera-

bility, that is, the perception and treatment of African students was another economic side effect:

> They are heading towards the First World, and we are like left behind. This, it's not their fault, but like we still have something to be recognized [for]. We have a culture, we have, we belong to a culture. (The Four Graduates)

While this student focused on the negation of African cultures, another student directly attributed the negation of individual African students to economic changes:

> The most recent Chinese economic reform/ economic miracle is creating a big problem in how they treat African students.[Because] Africa [is] an economically left behind continent, African students are considered globally, like by a Chinese national moral,. . . from a poor continent, then not an important person — regardless of what is meant in political speeches by Chinese officials in Africa. (Survey Note)

In a succinct statement, one Counsellor corroborated these observations. He stated that for African students, life in Chinese society was getting progressively more difficult:

> Before it was a question of only colour. . . . Now it is a question of colour and money. (Embassy Counsellor B)

The questions of "colour and money" and their overall impact on the experience of African students in China may be further considered by reflecting back upon the two approaches to international relations examined in Chapters Two and Three of this study.

Notes

.[1] *Rem min bi (rmb)* or *yuan* is the standard monetary unit in China.

.[2] One student itemized three points to express the degree to which he felt misled by "propaganda." He wrote: "1. Their level is not up to standard, yet still they want to fool the world that they can sponsor foreign students. 2. They use foreign students as scapegoats. They unknowingly take [sic] foreign students in decent places like hotels and take videos of these places as their living places. Then these video cassettes are sent to various countries to fool the people that foreign students are living in absolute better conditions. 3. They tell our Governments that students allowance is far above the salary of proffessors [sic]." A second student also expressed similar sentiments of feeling misled and scapegoated: "Foreign students are used to portray their [China's] false image abroad. They requested for large numbers of foreign students when they cannot cater well for them." A third student attached a separate piece of paper to the questionnaire to add, "After all these long years in China, the overall feeling is [that it has been] kind of a waste of time, a kind of victim[izaton] offered between two governments willing to keep up the political and cultural relations in . . . political speech[es]. . . . The spirit of cooperation as a whole is a failure because the impression I have got is that the African generation 'made in China' is not happy in general with the Chinese experience at all." (Notes from three survey, students from three countries)

In an interview, one student took these concerns to another level. He stated, "We are only like political hostages, just hostages by, made [by] contracts between the African governments and Chinese government, [and] for [their] interest [the contract] is called cultural exchange. But, when they make the deal [and] they send a number of student here, nothing, nothing is bridging" (The Four Graduates).

.[3] When discussing these bilateral relations, two Counsellors and one group of students emphasized a second point; that is, if an African country fell out of favourable relations with China, all students from that country would be sent home immediately, regardless of where the individual student may be in their degree program.

.[4] A number of writers have reported on this same phenomena. Hevi (1963: 185) revealed that "Whereas all other foreigners took 100 yuan a month each, the Albanians took 150. . . . There was nothing at all to warrant the Albanians taking more than we. . . . If the Albanians needed 150 yuan to live comfortably, we saw no reason why we should not do the same. This was one of the ways in which the Chinese tacitly acknowledged Albanian superiority over Africans." A few years later, Chen (1965: 117) reported the discrepancy became even greater: "Students from Africa and Asia, get one hundred *yuan*, and European and Soviet students received the generous amount of two hundred and fifty *yuan*." Chen explained that "One reason why European students are given a larger allowance is that they are served Western food instead of Chinese food." Goldman's (1965: 136) observations from the same time period concur. Though not referring specifically to African students, she wrote, "Chinese policies toward the foreign students were not uniform and there was a different attitude." She noted a "subtle gradation of preference, shifting with the mood of China's foreign policies, added further differentiation. . . .The graduation was also reflected in the amount of the scholarship awarded and the kinds of privileges accorded."

Burundian student, Barbatus Baryumyeko Gatoto graduates from the Faculty of Textile Engineering, Spinning and Weaving Department, China Textile University in Shanghai, 1993.

中国纺织大学特织89/班毕业留念 93.7

China Textile Univerdsity in Shanghai, Class of 1993. Foreign students from top left to right: Barbatus Gatoto from Burundi, Jeetendra Rhadka from Nepal, Medessi Dagan Maurice from Benin, Ronoh Cheruiyot Birgen from Kenya, and Julius Mwangi from Kenya.

CHAPTER TWELVE

Conclusions

The experience of African students in China today may be better understood by returning to the two approaches to international relations laid out at the start of this investigation: the Maoist view and the view of Galtung and Mazrui of the World Order Models Project. Linking these two sets of views provides an evaluative framework in which to place the findings and reflect upon the transformative possibilities of Sino-African exchanges within international relations and international knowledge relations.

CHINA'S CHANGING WORLD VIEW

In Chapter Two, the three main foreign policies of the Maoist reign served as a historical context in which to explore the evolution of China's definition of its place in the world. China's definition of its place, crystallized in Mao's Three-Worlds Theory, also provided a frame of reference for assessing any change or continuity in post-Mao global policy. The experience of African students in China may thus be seen as a reflection of the evolution of China's world view and a gauge of foreign policy orientations.

African Students in Maoist China

When African students first arrived in Beijing, they brought with them a Maoist image of the world that embodied the spirit of Bandung and an emerging revolutionary zeal which symbolically linked China's fate with the Third World and heralded the nations of Asia, Africa, and Latin America as a revolutionary force that would unite to end the dominance of the developed world and transform existing international order (Sautman, 1994). While this projection of a new world order had a strong appeal, most African students in Maoist China found the vision to have little grounding in reality. John Hevi's (1962) account of 118 African students in China between 1961-62 reveals that disillusionment and discontent grew so massively and so rapidly that within nine months, ninety percent of students (96 of the 118) returned to Africa.[1] Hevi (1962: 119-143) details the factors that caused this mass exodus: undesirable political indoctrination, language difficulty, poor educational standards, restrictions on social life, general hostility, spying, and racial discrimination. Hevi (1962: 183) identifies racial discrimination as "the first item on our list of grievances," but at the same time, this item appears to have been one among many factors. Moreover, financial hardship, though present, was not debilitating.[2] Thus, the essential elements of "colour and money," as the Counsellor put it, in the lives of African students today were not such exclusively overwhelming factors as they later became and remained.

This environment may be attributed to a Maoist view of the world that was marked by an ideological inversion: racial and social hierarchies were stratified in a thesis that extolled the virtues of 'coloured' people and the poor (Sautman, 1994). These ideals were featured in posters of Third World revolutionaries and in photos of Mao surrounded by exchange students of all races (Sautman, 1994: 428). Racial solidarity was particularly highlighted in Mao's pronouncements on anti-colonial and revolutionary movements in Africa and the black diaspora. At a time when the Chinese government officially justified its African aid projects on grounds of racial solidarity and the Red Guards rallied to support African causes, few would have dared to openly express hostility to people from the Third World (Sautman, 1994: 414). Yet with the death of Mao Zedong (1976) and the ascension of Deng Xiaoping (1978), China's world view changed.

African Students in the Reform Era (1978-present)

In December 1978, the Third Plenum of the Eleventh Central Committee of the Communist Party initiated the post-Mao era of reform, heralded the rise of Deng Xiaoping, and signified the acceptance of his open door policies (Hayhoe, 1989b; Sautman, 1994). The open door policies shifted attention from the promotion of international solidarity to the promotion of joint ventures, leases, investments, and trade with advanced capitalist nations. While these social and political shifts brought about tremendous openness and political reforms, they also resulted in a reemergence of social stratification and a rejection of Third World solidarity. As the preeminence of the poor was quickly replaced by the preeminence of national and individual enrichment, many began to feel humiliation — not solidarity — to be equated with the Third World (Sautman, 1994; Sullivan, 1994). Mao's Three-Worlds Theory, which symbolically linked China's fate with the Third World was quickly abandoned for a reform era vision which argued China's destiny lay with the West (Sullivan 1994: 443).

University Campuses in the Reform Era

This reform era vision was immediately manifested on university campuses. In July 1979, just seven months after the Third Plenum, a confrontation between Chinese and African students at the Shanghai Textile University set a pattern for a decade of conflict across campuses in Shanghai, Shenyang, Guangzhou, Beijing, Tianjin, Xian, Hangzhou, Wuhan, and Nanjing.[3] While many commentators focused on the motives behind the conflict, for others, these conflicts raised deeper concerns regarding the emerging world vision among China's future elites (Sautman, 1994: 423-429). Chinese students began to feel free to voice their support for the government's decision to cut interest free loans and technical assistance to the Third World but at the same time felt free to oppose the government's decision to continue the educational aid. Students felt that the government should not "waste" any more of its resources on others and particularity objected to the policy of "spoiling" African students while Chinese citizens suffered on lesser means,[4] in lesser conditions because of the failed economic policies of the Mao era (Sullivan 1994: 444). University administrations also began to react adversely to government policies that sent universities a quota of scholarship students but not the resources (Cheung, 1989). What the Chinese students and school authorities largely regarded as means to indulge, the African students regarded as means to control, isolate, alienate, and segregate them from the local community. Resentment built on all sides across many campuses.

Administrators' attempts to diffuse the tensions by dispersing African students across provincial universities only enlarged the problem (Sullivan, 1994: 444). Hostilities rose to such an extent that eventually Africans across China boycotted classes and ultimately demanded protection. In addition to security, the students insisted that authorities eradicate negative images of Africa through educational programs. University administrators considered their demands unwarranted and refused (Sautman, 1994; Sullivan, 1994).

Clashes of 1988-89

In 1988-89, these problems caught international attention when they culminated in the Nanjing Anti-African protests.[5] For a week long demonstration, 3,000 Chinese students marched in the streets, chanting antiblack, human rights, pro-democratic, and nationalistic slogans.[6] Moral indignation, sparked by racially motivated rumours, led Nanjing students to 'take the law into their own hands,' and just four months later this indignation and determination, combined with patriotism and anti-government sentiment, erupted in the most serious challenge to the CCP's rule since 1949 (Sullivan, 1994: 456-457). From this view, many contend that the Nanjing Anti- African protests heralded the pro-democracy movement of

1989,[7] as Chinese "democrats" used the anti-African sentiments to direct protest against the party regime" (Sautman, 1994: 426).[8]

Since 1989 no protests either against the party regime or against African students have been publicly staged. This does not mean, however, that such oppositions have since dissipated. In fact, as the data in this study reveals, problems of racial hostility and social isolation continue to plague African students, as they have in various forms and intensities since the Maoist era. Now in the second decade of reform, these problems have been exacerbated by economic impoverishment at a time when, ironically, China has emerged as the world's third largest economy and as America's largest trading partner after Japan.[9] The reform era view of the world, like the Maoist view of the world, has been marked by yet another ideological inversion; this time, however, racial and social hierarchies have been re-stratified by a world view which holds economic prosperity, in addition to race, as a key indicator of status. Thus, "colour and money," are the essential elements in the lives of African students today as China shifts its view of the world and its position in the world system. To gauge the shift of China's position in the international system, the World Order Models Project provides a holistic, global, and transformative paradigm.

WORLD ORDER MODELS PROJECT (WOMP)

The World Order Models Project, shaped by the goals of peace, economic well-being, social justice, and ecological balance, highlights the possibility for a nation's transformative role in international order. As discussed in Chapter Three, the four principles of Galtung's model of positive action (equity, autonomy, solidarity, participation) and Mazrui's three strategies of African modernization (domestication, diversification, counter penetration), provide evaluative conditions for determining the transformative possibilities of Sino-African relations within the global community.

Equity

Equity, according to Galtung, is characterized by mutuality; that is, mutual levels of planning and agreement on the aims and organization of the programs. In these Sino-African exchanges, degrees of equity are evident in that scholarships are established through bilateral agreements. However, the mutuality in the planning of these agreements and the mutuality of the interests served raises other issues. African students urged Chinese authorities to seek full input on an individual and national level; that is, ask individual students to identify their background (personal, academic, and professional), educational needs, and goals. Likewise, ask individual African countries to identify their background (national conditions and national priorities), educational needs, and goals. From here, students urged Chinese authorities to fully consider these individual and nationally identified circumstances. To best meet these circumstances, students felt African nations must be given the opportunity to play a full role in the design and definition of these programs.

Efforts to engage Africans at the individual and national level will not only increase the mutuality of the planning but may also reduce students' problems of obtaining full and correct information. Inadequate and inaccurate prior information and ensuing feelings of being misled, particularly in regards to the language of instruction, years of program, and sufficiency of the stipend, may be alleviated if terms of agreement were reached through more mutual and equitable means.

Autonomy

While students arrive with little knowledge about China, their hosts also know little about Africa. If Chinese educational authorities plan these programs without full consultation and without a 'reasonable' knowledge of African conditions, autonomy, as defined by Galtung, will remain elusive. Autonomy requires centre participants to have a knowledge of the periphery participants in order that theoretical perspectives may be appropriately rooted in their culture. Cultural ignorance not only precludes autonomy, but it also inhibits a sense of solidarity.

Solidarity

Solidarity, according to Galtung, suggests forms of knowledge transfer that encourage maximum interaction and linkages between all participants that will evolve into a collective reinterpretation and broad dissemination of the new knowledge. In China, however, African students reported that new knowledge is not freely disseminated. Access to appropriate technology, such as rice harvesting, and access to sites, such as local industries and manufacturing plants, was denied and knowledge withheld. Students felt knowledge was competitively guarded as a source of national property.[10] Thus, while Mazrui frames the problem of African nations needing to "permit" non-Western civilizations to "reveal their secrets," in this case the issue is not so much Africa's willingness to permit but rather China's willingness to reveal.

Moreover, students reported that any attempts to 'reveal' African 'secrets,' that is, to reverse the flow of influence, was regarded as suspect. Mazrui's notion of "counter penetration," the dissemination of knowledge *from* Africa, is marred by institutional boundaries that segregate cultural and intellectual integration. Forms of organization not only discourage but also prohibit full interaction. For foreign students, the scarcity of local attendance at their seminars signifies that forms of knowledge, values, and concepts from Africa are neither fully welcome nor duly honoured in China. Students' attempts to both contribute and receive knowledge are met with resistance. Thus, in Sino-African exchanges, Mazrui's condition of diversification and "full reciprocal international penetration" and Galtung's condition of solidarity have not been fully realized.

There may be two consolations. One, the domestication of knowledge, as defined by Mazrui, while left entirely in the hands of individual students, may be enabled by African-only seminars and classes. African-only situations may allow students to collectively focus and concentrate on reinterpreting and transforming the knowledge to make it relevant to a particular African context. While students researching "purely Chinese based" topics, such as rapeseed and soya bean, reported little hope of domesticating this knowledge, others, such as the student researching fruit preservation and vegetable storage, felt this new knowledge could be easily domesticated and applied either alone or as supplements to indigenous knowledge of his country. Two, the dissemination of the new knowledge in the African context may be broad because the problem of 'brain drain', often found in North-South relations, is less likely to occur in this situation. However, it bears emphasis that these two points are strictly consolations. Above all else, African students and the three Embassy Counsellors on their behalf expressed deep regret about the lack of solidarity, integration, and participation on Chinese campuses.

Participation

Participation, in Galtung's view, is an approach to knowledge transfer that does not marginalize or stratify in a hierarchical way but rather elicits the creative contributions of the peripheral scholar. On one hand, China's approach to scholarship offerings and language acquisition supports wide participation. In fact, in all the world, China may be the country which elicits the most participation from African scholars. As the three Counsellors remarked, to their knowledge, no other country regularly and reliably offers African nations ten, cumulative, full scholarships per year.[11] Moreover, these scholarships are publicly advertised and often awarded by meritocratic means. And while these means do seem to draw participation of students from diverse socioeconomic backgrounds, they do not reach women and men to the same degree. The 14.3% of women in this study may even be proportionally larger than their overall presence in these programs. In regards to recruiting the participation of women, to date, Sino-African programs have no special initiative.

However, China's approach to language acquisition shows strong initiative and supports their wide invitation for participation. For undergraduates, the approach to language is affirmative and not used as a means to restrict access. It is a given that after one year of training, students will acquire sufficient language skills to begin their degrees and in the course of their studies, students' proficiency will necessarily improve. At the end of the five years, students will be fluent in Chinese and have a degree in hand. This approach provides an interesting contrast and positive alternative to the approaches of the Test of English as a Foreign Language (TOEFL), for example, where access to degree programs is restricted until students demonstrate very high levels of fluency.

While the generous approach to scholarship offerings and language acquisition may affect the number of African participants, it does not, however, increase their actual level of engaged participation on campus. Students expressed profound disappointment with the narrowness of the intellectual or social contributions expected, wanted, or accepted of them. Students' hopes of participating as colleagues, teaching assistants, research assistants, or in some cases as part-time workers, had not been realized. Students' hopes of being integrated into school activities, such as cultural programmes, entertainment, and sports, had also not been fulfilled. Students felt this deliberate and active negation of their involvement, through institutional, societal, and economic barriers, marginalized their full participation and exacerbated cultural ignorance and racial antagonism.

FINAL COMMENTS

In the post-Mao era of reform, Galtung's conditions of equity, autonomy, solidarity, and participation, and Mazrui's conditions of domestication, diversification, and counter penetration have not been fully realized. Although these shortcomings may distract, they need not diminish the accomplishments and future possibilities for Sino-African relations. Both governments and individuals in China and African nations consider these exchanges to be a source of considerable value and have gone to great lengths to secure opportunities to train, be trained, and thus sustain Sino-African relations. Sustained relations for China represent a continued strategic opportunity. For many African nations, sustained relations with China represent an opportunity to begin to realize what for many is a political priority; that is, to overcome the colonial traces in the content and substance of their educational experiences (Weiler, 1984: 188). A decolonized pedagogical paradigm offers an opportunity to ensure, in the words of Amadou Mahtar M'Bow, "the full development of cultural identity" (M'Bow, Senegalese educator and former director-general of UNESCO, as cited by Weiler, 1984: 192).

China, as a non-colonized Third World society, with an indigenous modern educational system and an independent socialist economic system,[12] holds the potential to partake in this opportunity and make a significant contribution to an alternative pedagogy and an alterative rethinking of international knowledge transfer. However, as the data in this study reveals, this opportunity has not been fully embraced and thus this potential has not been fully realized. Similar to the experience of other Third Word nations, China's integration into the capitalist world order underscores the power of the dominant mechanisms that maintain the international status quo (Hayhoe, 1989: 97). This reality highlights what Weiler (1984: 189) considers to be part of the most critical aspects of "underdevelopment": the dominate mechanisms of economic, cultural, scientific, and professional control that have been generated and sustained, in part, by systems of knowledge production and higher education in the centre countries. Though China's self-reliant economy and strong socialist institutions make it less vulnerable to economic dependency, cultural alienation, and social divisiveness threatening many other peripheral, postcolonial societies of the Third World, its integration into the capitalist world order is nevertheless similar to the experience of other Third Word nations in that China currently plays more of a supporting than transforming role in international power relations (Hayhoe, 1986b; 1989). This notwithstanding, WOMP scholarship offers the optimistic hope for emerging signs and action strategies that promote structural transformation to greater equality. Such emerging signs and action strategies may be revealed through future investigations.

Future Prospects

The purpose of this investigation, set out in Chapter One, was to acknowledge the sustained educational cooperation between China and Africa and place this knowledge within a larger literature on approaches to international and academic relations, which to date has focused almost exclusively on the perspective of North-South relations. The limitations of this study and the paucity of literature on South-South educational relations raise a plethora of possibilities for future studies. Future investigations of Sino-African exchanges pursued, for example, by a scholar from China or Africa may overcome many of the significant limitations of this study. My location in this study, partially defined by culture, race, and gender, necessarily limits my perspective. A Chinese or African scholar may offer a perspective with insights and interpretations considerably different from mine. Such scholars directly involved in these exchanges would add the further dimension of lived experience. Moreover, this investigation conducted in the Chinese language would access the cultural context of this setting to an infinitely greater degree than I have been able to. Other exciting research possibilities emerge by moving the cultural context of this investigation from China to one or more African nations and thereby shifting the fundamental premise for the entire project. While the findings of this investigation are largely the voices of African students, this study has been anchored in a historical body of literature related to a Chinese world view. Future studies anchored in a body of literature related to a world view and foreign policy of an African nation would surely add rich and contrasting dimensions to this investigation. Such a study might be further informed by a theoretical body of literature on international academic relations rooted in perspectives of African scholars. Issues of global equity, race, difference, and the decolonization of knowledge in the international arena are a few areas of possible focus for such investigations. Another direction for future studies could be gender: specifically the participation, or lack thereof, of women, an aspect of these exchanges barely touched upon here. One may choose to add a comparative dimension, for example, contrasting the experience of men and women. In fact, comparative possibilities seem almost endless. For example, one may compare this exchange with other South-South exchanges such as those offered in the Phillippines, India, Mexico, Argentina, and Egypt to which students from the South tend to flow (Altbach et al., 1985). One may focus on these or other South-South programs within China, within Africa, or beyond. And of course, one may choose to compare Southern and Northern experiences.

Finally, a possibility which occurred to me throughout the course of the data collection would be a study devoted to the community of African students who remain in China years after their graduation. Most of these students cannot go home, often because of civil conflict, and cannot get a visa to another country. In a sense, they are stranded in that the Chinese government permits them to stay but does not permit them to work. In other words, they are allowed to live, but they are not allowed to make a living in China.[13] They are thus obliged to be fully supported by the United Nations High Commission for Refugees (UNHCR). Members of this community, many of whom still live in university dormitories throughout the major urban centres of Beijing, Guangzhou, and Shanghai, would provide a rich area of study in a variety of subject areas.

The possibilities that I have suggested are just a few of the many which I hope this study has raised. Yet while this study may point to a variety of options for future studies, students overwhelmingly identified a single priority. Remaining true to the data, I end by highlighting that in this study a total of 133 undergraduates, post graduates, and graduates from twenty-nine African nations, pursing degrees in more than twelve disciplines, in fourteen sites, across four cities, spoke volumes in a remarkably consistent voice: colour and money, above all else, were the essential elements in their lives. While this study may reveal these elements, it does not begin to heed the call for a full systematic investigation into the discourse of race and class in China.

In order to assist those who may pursue further studies in this area, I would like to offer a look at a few scholars who have begun to pave the way. The June 1994 edition of *The China Quarterly* has a section devoted to a 'Focus of Race and Racism in China'. In this final chapter, I have made much use of the three articles in this section written by Dikötter, Sautman, and Sullivan. These scholars, like the students in this study, urge others to recognize that reform era attitudes towards race and class have far-reaching significance that to date has been largely ignored. Dikötter (1994: 403) points out that while a considerable body of scholarly work has revealed much about the historical and contemporary dimensions of racial identities in the West, comparatively little is known about racial identities elsewhere, and "virtually nothing is known about the articulation and deployment of racial frames of reference in China." Dikötter (1994: 405-412) argues that racial identities in the modern world are framed in exclusively Western terms which marginalize and trivialize other discourses.[14]

Thus, the current parameters which tend to frame the discourse of race relations into a North-South phenomenon, like the current parameters which tend to frame the discourse of international knowledge relations into a North-South phenomenon, marginalize South-South experiences. The challenge, it seems, is to search for broader, more inclusive understandings in both fields. Thus, this study closes with an acknowledgement of the considerable work yet to be done. In attempting to draw attention to a South-South aspect of knowledge relations, the participants in this study first insisted that attention be drawn towards a South-South aspect of race relations. Expanding the field of international knowledge relations and race relations to consider a South-South dimension would surely illuminate signs and action strategies that would promote WOMP's optimism for positive structural transformations within the global community. The complex intertwining of social status, race, and class affects the transfer of knowledge, in all directions, North and South.

Notes

.¹ Reactions to this study have been mixed and at times emotional. Snow (1988: 199) judges the text to be "savagely polemical," and characterizes Hevi as "something of a toady, eager to win the favour of the Institute authorities, but was soured by his inability to find a girlfriend." Snow portrays Hevi as a man "spoiling for revenge, and that his rancour was exploited by American officials." Sautman (1994: 414) refers to the text as a "hostile account," and Larkin (1971: 142) and Hutchison (1975: 186) describe the text as "somewhat polemical," but both add that the allegations have never been disproved, and Hutchison points out that actually the allegations "were substantiated by Zambian students who underwent similar experiences ten years later, in 1972." Sautman (1994: 414) offers a few details of these experiences stating that "in 1970 Africans returned to China but by 1972 became so discontent that they deliberately burned portraits of Mao Zedong so that they would be deported." The reports from African students in China today, as revealed by the findings in this study, share similar key points and would tend to support Hevi's account.

.² Hevi (1962: 113-114) reports that in the fall term of the 1960-1961 academic year, the allowance for African students began at 80 *yuan* per month and was raised to 100 *yuan* per month by the second term. By contrast, Chinese students received 10 *yuan* and graduate teachers received 40 *yuan* each month. Hevi notes that "To the Chinese we must have looked a really ungrateful lot. . . . But the fact is that we simply could not live on the starvation ration offered to Chinese students and tutors."

.³ The *Beijing Review* (1987 01 19, 01 26), Cheung (1989), Delfs (1989), Scott (1986 06 19; 06 26), and Xia (1989) are a few among many who reported on the Chinese-African campus confrontations between 1978 and early 1989. Various commentators attribute the conflicts to a disparity in standards of living, antagonism over interracial dating, xenophobia, and racism. The Chinese government dismisses all of the above as "rumour mongering" and contends that these skirmishes were provoked by only a few Chinese students who shouted derogatory remarks at a few disruptive African students. Crane (1994), Dikotter (1994), Sautman (1994), and Sullivan (1994) provide detailed examinations of these incidents and discuss their far reaching significance. Sautman (1994: 415-423) reports on the earlier conflicts while Crane (1994) and Sullivan (1994) focus on the particulars of the week long violence in Nanjing, 1989.

.⁴ The disparity over the living conditions was noted by Scott (1986b: 51) who reported that by 1986 the Chinese Government provided African students with a monthly stipend of 180 *yuan* while Chinese students received 23 *yuan* the per month.

.⁵ The attention and "negative commentary" from government and opinion leaders from Kenya, Liberia, Gambia, Ghana, Libya, Benin, Senegal is detailed by Sautman (1994: 422).

.⁶ Sautman (1994: 420-423) reports that though the week in Nanjing was the largest and most sustained protest, shortly after this incident, anti-African protests broke out in Hangzhou, Beijing, and Wuhan.

.⁷ Sautman (1994: 426) points out that student "democrats" linked human rights and freedom to anti-black slogans and thus advanced their charge that the regime had failed to protect the rights of Chinese citizens against alleged crimes of the Africans. The link between universal rights and racial antagonism is incongruous but not unprecedented as seen before in Bosnia, India, South Africa, and United States, for example.

.⁸ The week long demonstrations left African students demanding to be returned home because of "entrenched racial discrimination, police brutality, segregation, and 'a lack of human decency' towards them" (Cheung, 1989: 32). African students felt they were used as scapegoats by the demonstrators, on the one hand, who used anti-African sentiment to protest the government, and by the Chinese government, on the other hand, who manipulated anti-African sentiment to channel and divert the protest away from the party (Sautman, 1994: 426). (Delfs, 1989: 12) states that "What is particularly tragic is that the Chinese mob's reaction to the initial incident was unnecessarily aggravated by the authorities. Their attempts to suppress accurate reports of the original incident and subsequent Chinese demonstrations, and their failure to counter misleading rumours, helped to inflame mob sentiment against African students."

.⁹ Theroux (1993: 34) reports that "China has emerged as the world's third largest economy, according to the International Monetary fund, and as America's largest trading partner after Japan. In the first quarter of 1993, China's GNP grew at an annual rate of 14 percent, outstripping every country in the world; in contrast, America grew, in the first quarter of this year, at a 1 percent annual rate. The engine driving most of China's growth is centered in the southern provinces, a region of 290 million people, where the government has established five Special Economic Zones."

.¹⁰ One student believed that knowledge was withheld from African students because of potential competition between "you know, guys from the Third World" (Interview, Post Graduate A). However, 'Third World guys' may not be seen as the only source of competition. Hayhoe (1989b: 97) reports that internal dimensions within Chinese scholarly communities are such that "new knowledge is not freely disseminated to other institutions or the local community; rather, it is jealousy guarded as a source of prestige and power by the particular institution that is its possessor. The transformation of the structure and organization of knowledge in the higher curriculum is thus more cosmetic than real, with new knowledge placed in persisting patterns of hierarchy and control."

.¹¹ The Counsellors' observations are backed up by Kanduru (1997: 178) who reports that in the case of Tanzania, "the one that donates the greatest number of scholarships is China. The Government of China donates about 30 scholarships a year."

.¹² Hayhoe (1986b: 534) points out the features which distinguish China from many Third World countries. One, China is not a postcolonial society. Two, China's modern educational system, while incorporating select foreign educational patterns, was created indigenously, preserving certain cohesive Confucian and Maoist values. Three, China has been a socialist country since 1949. From that time, the Chinese economy has been largely independent of the advanced capitalist world, and from 1960 it has been independent from the Soviet bloc.

.[13] This point was repeatedly emphasized by the graduates I met across China. These graduates stressed that within the large foreign communities in China, Westerners and other Asians, who, for the most part, have not been educated in China, do not speak fluent Chinese, and do not have the years of cultural exposure, can readily find employment. But for Africans, educated in China, fluent in Chinese, and with years of living experience "there is no way [to] get [work]. So .. the question is always, why? ... The education we got, the science we got, it is all from them [China]. And we, you know, [speak] French, English, Chinese, have Master's degrees. . . [but we] will never get a job! Why?" (The Four Graduates). This question and the multitude of issues it raises fell outside the parameters of this study but was vital to those graduates I met.

.[14] According to Dikötter (1994: 405-412), the discourse of race in China cannot be minimized as a consequence of the "hegemonic powers" of Euro-American imperialism for three reasons. First, it perpetuates a unitary conception of racism which is universal in its origins (the West), its causes (capitalist society), and its effects (colonialization). Second, it disregards the "historical specificities" of racial identities and reduces a variety of cultural groups into the West and the Rest. Third, it represents people in China as a passive subject devoid of free thought, critical analysis, and intellectual autonomy. The "western impact-Chinese response" approach imposes "eurocentric" distortions and negates historical transformations which occurred in China long before its prolonged exposure to foreign thoughts (Dikötter, 1994: 409). Dikotter's (1991) *The Discourse of Race in Modern China* is perhaps the most systematic investigation of the topic to date.

References

Altbach, P. (1977). Servitude of the mind? Education, dependency, and neocolonialism. *Teachers College Record, 79,* 187-204.

Altbach, P. (1980). The distribution of knowledge in the Third World: A case study in neocolonialism. In P. G. Altbach & G. Kelly (Eds.), *Education and colonialism* (pp. 302- 330). New York: Longman.

Altbach, P. G., Kelly, D. H., & Lulat, Y. G.-M. (1985). *Research on foreign students and international study: An overview and bibliography.* New York: Praeger.

Arnove, R. (1980, February). Comparative Education and World Systems Analysis. *Comparative Education Review, 24,* 48-62.

Ashbury, F. D. (1991). *International scholarly exchange and status recognition: A case study of China's exchange scholars and students in Ontario universities.* Unpublished doctoral dissertation, York University, Toronto, Ontario.

Babbie, E. R. (1973). *Survey research methods.* California: Wadsworth.

Bailey, C. A. (1996). *A guide to field research.* California: Pine Forge Press.

Barber, E. G., Altbach P. G., & Meyers, R. G. (Eds.). (1984). *Bridges to knowledge: Foreign students in comparative perspective.* Chicago: The University of Chicago Press.

Barnett, G. A., & Wu, R.Y. (1995). The international student exchange network: 1970 & 1989. *Higher Education, 30,* 353-368.

Bastid, M., & Hayhoe, R. (Eds.). (1987). *China's education and the industrialized world.* New York: M. E. Sharpe, Inc.

Bates, R. H., Mudimbe, V. Y., & O'Barr, J. (Eds.). (1993). *Africa and the disciplines: The contributions of research in Africa to the Social Sciences and Humanities.* Chicago: University of Chicago Press.

Bereday, G. (1964). *Comparative method in education.* New York: Holt, Rinehart and Winston.

Boonyawiroj, S. (1982). *Adjustment of foreign graduate students: Nine case studies.* Unpublished doctoral dissertation, University of Toronto, Ontario.

Borman, K. M., Goetz, J. P., & LeCompte, M. D. (1986). Ethnographic and qualitative research and why it doesn't work. *American Behavioural Scientist, 30* (1), 42-57.

Burgen, A. (Ed.). (1996). *Goals and purposes of higher education in the 21st century.* London: Jessica Kingsley Publishers.

Chen, T. H. (1965). Government encouragement and control of international education in communist China. In S. Fraser (Ed.), *Governmental policy and international education* (pp. 111-133). New York: John Willey and Sons, Inc.

Chen, T. H. (1974). *The Maoist educational revolution.* New York: Praeger.

Chen, T. H. (1981). *Chinese education since 1949: Academic and revolutionary models.* New York: Pergamon Press.

Cheung, T. M. (1989, February 16). Frayed welcome mat. *Far Eastern Economic Review, 32.*

Chinese African people's friendship association. (1961). Peking: Foreign Language Press.

Chinese people resolutely support the just struggle of the African people. (1961). Peking: Foreign Language Press.

Cooley, J. K. (1965). *East wind over Africa: Red China's African offensive.* New York: Walker and Company.

Crane, G. T. (1994, July). Collective identity, symbolic mobilization, and student protest in Nanjing, China, 1988-1989. *Collective Politics, 26,* 395-412.

Curtin, P. D. (1981). Recent trends in African historiography and their contribution to history in general. In J. Ki-Zerbo (Ed.), *General history of Africa I: Methodology and African prehistory* (pp. 54-71). Paris: United Nations Educational, Scientific and Cultural Organization.

Davies, B. (1982). *Life in the classroom and playground: The accounts of primary school children.* London: Routledge and Kegan Paul.

Delamont, S. (1992). Beauty lives though lilies die: Analysing and theorizing. In *Fieldwork in educational settings: Methods, pitfalls and perspectives* (pp. 149-162). London: The Falmer Press.

Delfs, R. (1989, January 12). Racist mob mentality: Chinese students stone black African colleagues. *Far Eastern Economic Review,* 12.

Dikötter, F. (1992). *The discourse of race in modern China.* London: Hurst & Company.

Dikötter, F. (1994, June). Racial identities in China: Context and meaning. *The China Quarterly, 138,* 404- 410.

Duyvendak, J. J. L. (1949). *China's discovery of Africa: Lectures given at the University of London on January 22 and 23, 1947.* London: Arthur Probsthain.

Eisner, E. W. (1991). *The enlightened eye: Qualitative inquiry and the enhancement of educational practice.* New York: Macmillan Publishing.

Ekaiko, Uko, T. C. (1981). *The effect of selected cultural and environmental factors on the social and academic adjustments of African students in the United States universities.* Unpublished doctoral dissertation, Wayne State University, Detroit, Michigan.

Ellis, C., & Flaherty, M. (1992).An agenda for the interpretation of lived experience. In C. Ellis & M. G. Flaherty (Eds.), *Investigating subjectivity: Research on lived experience* (pp. 1-13). Newbury Parks, CA: Sage Publications.

Fabricated letter arouses concern. (1987, January 19). *Beijing Review, 03,* 8.

Fage, J. D. (1978). *A history of Africa.* London: Hutchinson & Co (Publishers) Ltd.

Falk, R. (1982). Contending approaches to world order. In R. Falk, S. S. Kim, & S. Mendlovitz (Eds.), *Towards a just world,* (Vol.1). Boulder, Colorado: Westview Press.

Faust J. R., & Kornberg, J. F. (1995). *China in world politics*. Boulder: Lynne Reinner Publishers.

Feiz, Y. (1995). *A study of problems faced by post-graduate visa students at the University of Toronto and York University*. Unpublished doctoral dissertation, University of Toronto, Ontario.

Filesi, T. (1972). *China and Africa in the Middle Ages*. London: Frank Cass.

Finch, J. (1988). Ethnography and public policy. In A. Pollard, J. Purvis, & G. Walford (Eds.), *Education, training and the new vocationalism: Experience and policy* (pp. 185-200). Milton Keynes: Open University Press.

Friendship remains strong forever. (1987, January 26). *Beijing Review, 04*, 28.

Galtung, J. (1975). Is peaceful research possible? In *Peace: Research. Education. Action: Essays in peace research* (pp. 263-279). Copenhagen: Christian Eglers.

Goldman, R. (1965). The experience of foreign students in China. In S. Fraser (Ed.), *Governmental policy and international education* (pp. 135-140). New York: John Willey and Sons, Inc.

Gray, G., & Guppy, N. (1994). *Successful surveys: research methods and practice*. Toronto: Harcourt Brace & Co.

Hammersley, M., & Atkinson, P. (1983). *Ethnography: principles in practice*. New York: Tavistock Publications.

Hans, N. (1971). *Comparative Education*. London: Routledge & Kegan Paul.

Hao, Y., & Huan, G. (Eds.). (1989). *The Chinese view of the world*. New York: Pantheon Books.

Hawkins, J. N. (1974). *Mao Tse-Tung and education: His thoughts and teachings*. Connecticut: Linnet Books By the Shoe String Press, Inc.

Hay, J., (Ed.). (1995). *Boundaries in China*. London: Reaktion Books Ltd.

Hayhoe, R. (1986a). China, Comparative Education and the World Order Models Project. *Compare: A Journal of Comparative Education, 16* (1), 65-80.

Hayhoe, R. (1986b). Penetration or mutuality? China's educational cooperation with Europe, Japan, and North America. *Comparative Education* Review, *30* (4), 532-559.

Hayhoe, R. (1989a). A Chinese puzzle. *Comparative Education Review, 33* (2), 155-175.

Hayhoe, R. (1989b). *China's universities and the open door.* New York: M. E. Sharpe, Inc.

Hayhoe, R. (Ed.). (1992). *Education and modernization: The Chinese experience.* Oxford: Pergamon Press.

Henry, N. (1976, December). An African in Peking. *Crisis, 83*, 339-344.

Hevi, E. J. (1963). *An African student in China.* New York: Praeger.

Hevi, E. J. (1967). *The dragons embrace.* New York: Praeger.

Hickey, D., & Wylie, K. C. (1993). *An enchanting darkness: The American vision of Africa in the twentieth century.* East Lansing: Michigan State University Press.

Holmes, B. (1965). *Problems in education: A comparative approach.* London: Routledge & Kegan Paul.

Holmes, B. (1981). *Comparative Education: Some considerations of method.* London: George Allen and Unwin.

Holsti, K. J. (1985). *The dividing discipline: Hegemony and diversity in international theory.* Winchester, Mass.: Allen & Unwin.

hooks, b. (1988). *Talking back: Thinking feminist, thinking black.* Toronto: Between the Lines.

Huang, C. S. (1994). *International scholars as participants in American international educational exchange.* Unpublished doctoral dissertation, University of Miami, Florida.

Hutchison, A. (1975). *China's African revolution.* Great Britain: Hutchinson & Co (Publishers) Ltd.

Jackson, N. (1991). Changing the subject: A voice from the foundations. *McGill Journal of Education, 26* (2), 125-136.

Jacobsen, C. K. (1993). *Facilitators of international educational programs in China.* Unpublished doctoral dissertation, Graduate Department of Education, University of Toronto, Ontario.

Ji, K. (1993). *A study and comparative analysis of international student needs at Mississippi State University.* Unpublished doctoral dissertation, Department of Educational Leadership, Mississippi State University.

Jobe, J. C. (1994). *Characteristics of established international student programs at three Florida universities.* Unpublished doctoral dissertation, The Florida State University College of Education.

Joron, M. F. (1992). *Women and personal meaning of literacy: Beyond the mere reading and writing.* Unpublished monograph, Faculty of Education, McGill University, Montreal, Quebec.

Kandel, I. (1955). *The new era in education.* Cambridge, Mass.: Houghton and Mifflin, Inc.

Kanduru, A. I. (1997). *The implementation of the national manpower policy by Tanzanian universities from 1962 to 1994.* Unpublished doctoral dissertation, Faculty of Education, University of Toronto, Ontario.

Kim, S. (1979). *China, the United Nations, and World Order.* New Jersey: Princeton University Press.

Kim, S. (1980). Mao Zedong and China's changing world view. In J. C. Hsiung & S. S. Kim (Eds.), *China in the global community* (pp. 16-39). New York: Praeger.

Kim, S. (1984). China and the Third World: In search of a neorealist world policy. In S. S. Kim (Ed.), *China and the world: Chinese foreign policy in the post-Mao era* (pp. 178-211). Boulder: Westview Press.

Kim, S. (1994). China and the Third World in the changing world order. In S. S. Kim (Ed.), *China and the world: Chinese foreign relations in the post-Cold War era* (pp. 128-168). Boulder: Westview Press.

Ki-Zerbo, J. (1981). General introduction. In J. Ki-Zerbo (Ed.), *General history of Africa I: Methodology and African prehistory* (pp. 1-23). Paris: United Nations Educational, Scientific and Cultural Organization.

Klineberg, O., & Hull, W. F., IV. (1979). *At a Foreign University: An international study of adaptation and coping.* New York: Praeger.

Konyu-Fogel, G. (1994). *The academic adjustment of international students by country of origin at a land-grant university in the United States.* Unpublished doctoral dissertation, College of Human Resources and Education, West Virginia University.

Larkin, B. D. (1971). *China and Africa 1949- 1970: The foreign policy of the People's Republic of China.* Los Angeles: University of California Press.

Larkin, B. D. (1980). China and the Third World in global perspective. In J. C. Hsiung & S. S. Kim (Eds.), *China in the global community* (pp. 63-84). New York: Praeger.

Lauwerys, J. A. (1959). The philosophical approach to Comparative Education. *International Review of Education, 5* (3), 281-296.

Lawler, P. (1995). *A question of values: Johan Galtung's Peace Research.* Boulder, Colorado: Lynee Rienner.

Levine, S. I. (1980). The superpowers in Chinese global policy. In J. C. Hsiung & S. S. Kim (Eds.), *China in the global community* (pp. 40-62). New York: Praeger.

Levine, S.I. (1989). Chinese foreign policy in the strategic triangle. In J. T. Dreyer (Ed.) & I. J. Kim (Series Ed.), *Chinese defence and foreign policy* (pp. 63-86). New York: Paragon House.

Lin, Z. (1989). China's Third World policy. In Y. Hao & G. Huan (Eds.), *The Chinese view of the world* (pp. 225-259). New York: Pantheon Books.

Lulat, Y. G.-M. (1984). International students and study- abroad programs: A select bibliography. In E. G. Barber, P. G. Altbach, & R. G. Meyers (Eds.), *Bridges to knowledge: Foreign students in comparative perspective* (pp. 207-246). Chicago: The University of Chicago Press.

Mallinson, V. (1975). *An introduction to the study of Comparative Education.* London: Routledge and Kegan Paul.

Masemann, V. L. (1990). Ways of knowing: Implications for Comparative Education. *Comparative Education Review, 34* (4), 465-473.

Mazrui, A. A. (1975a, May). The African university as a multinational corporation: Problems of penetration and dependency. *Harvard Educational Review, 45* (2), 191-210.

Mazrui, A. A. (1975b). World culture and the search for human consensus. In S. Mendlovitz (Ed.), *On the creation of a just world order: Preferred worlds for the 1980/90s* (pp. 1- 37). New York: Free Press.

M'Bow, A.-M. (1981). Preface. In J. Ki-Zerbo (Ed.), *General history of Africa I: Methodology and African prehistory* (pp. xvii-xxi). Paris: United Nations Educational, Scientific and Cultural Organization.

McLean, M. (1983). Educational dependency: A critique. *Compare, 13* (1), 25-42.

McMahon, M. E. (1988). *Knowledge acquisition in the global market: Third World participation in international study.* Unpublished doctoral dissertation, Stanford University.

Mickle, K. M. (1984). *The cross cultural adaptation of Hong Kong students at two Ontario universities.* Unpublished doctoral dissertation, University of Toronto, Ontario.

Noah, H., & Eckstein, M. (1969). *Towards a science of Comparative Education.* London: MacMillan.

Ogunsanwo, A. (1974). *China's policy in Africa 1958-71.* Cambridge: Cambridge University Press.

O'Leary, G. (1980). *The shaping of Chinese foreign policy.* London: Croom Helm Ltd.

Pennycook, A. (1992). *The cultural politics of teaching English in the world.* Unpublished doctoral dissertation, Graduate Department of Education, University of Toronto, Ontario.

Poorshaghaghi, S. (1992). *A study of adjustment problems of international students at Northern Virginia Community College.* Unpublished doctoral dissertation. Faculty of the School of Education and Human Development, George Washington University.

Porter, J. (1962). *The development of an inventory to determine the problems of foreign students.* Unpublished doctoral dissertation, College of Education, Michigan State University.

Portères, R., & Barrau, J. (1981). Origins, development and expansion of agricultural techniques. In J. Ki-Zerbo (Ed.), *General history of Africa I: Methodology and African prehistory* (pp. 687-705). Paris: United Nations Educational, Scientific and Cultural Organization.

Salam, A. (1986). Inter-University co-operation: The Europe-Developing Countries experience: Balance sheet and perspectives. *Higher Education in Europe, 11* (3), 7-12.

Sautman, B. (1994, June). Anti-black racism in post-Mao China. *The China Quarterly, 38,* 413-437.

Schram, S. R. (1963). *The political thought of Mao Tse-tung.* New York: Praeger.

Scott, M. (1986a, June 19). Blacks and red faces: Allegations of racism embarrass the Chinese authorities. *Far Eastern Economic Review,* 20.

Scott, M. (1986b, June 26). Black students and the tide of prejudice. *The Far Eastern Economic Review,* 51.

Sheriff, N. M. (1981). Nubia before Napata. (-3100 to -750). In G. Mokhtar (Ed.), *General history of Africa II: Ancient civilizations of Africa* (pp. 245-297). Paris: United Nations Educational, Scientific and Cultural Organization.

Shih, C-Y. (1993). *China's just world: The morality of Chinese foreign policy.* Boulder: Lynne Rienner Publishers, Inc.

Smith, M. (1974). *The attitudes of new Canadian high school students and their achievement in English.* Unpublished master's thesis, University of Toronto, Ontario.

Snow, P. (1988). *The star raft: China's encounter with Africa.* Great Britain: The Bath Press.

Sono, T. (1993). *Japan and Africa: The evolution and nature of political and human bonds, 1543-1993.* Pretoria, South Africa: HSRC Publishers, Sigma Press (Pty) Ltd.

Spaulding, S., & Flack, M. J. (1976). *The world's students in the United States: A review and evaluation of research on foreign students.* New York: Praeger.

Sullivan, M. J. (1994, June). The 1988-1989 Nanjing anti-African protests: Racial nationalism or national racism? *The China Quarterly, 138,* 438-57.

Sutter, R. G. (1986). *Chinese foreign policy: Developments after Mao.* New York: Praeger.

Theroux, P. (1993, October). Going to see the dragon. *Harper's Magazine,* 3-56.

Tuso, H. (1981).*The academic experience of African graduate students at Michigan State University.* Unpublished doctoral dissertation, Department of Administration and Higher Education, Michigan State University.

UNESCO statistical yearbook. (1989). Paris: United Nations Educational, Scientific and Cultural Organization.

UNESCO statistical yearbook. (1995). Paris: United Nations Educational, Scientific and Cultural Organization.

VOA report called nonsense (1989, January, 16-22). *Beijing Review, 9.*

Vérin, P. (1981). Madagascar. In G. Mokhtar (Ed.), *General history of Africa II: Ancient civilizations of Africa* (pp. 693-717). Paris: United Nations Educational, Scientific and Cultural Organization.

Walton, F. C. (1985). *International training transfer: an investigation of the variables critical to the process.* Unpublished doctoral dissertation, University of Maryland.

Warwick, D. P., & Lininger, C. A. (1975). *The sample survey: theory and practice.* New York: McGraw-Hill.

Weber, M. (1949). *The methodology of the social sciences.* New York: The Free Press.

Weiler, H. (1984). The political dilemmas of foreign study. In E. G. Barber, P. G. Altbach, & R. G. Meyers (Eds.), *Bridges to knowledge: Foreign students in comparative perspective* (pp. 184-195). Chicago: The University of Chicago Press.

Weiler, H. (1989, June 28). *The mediation of knowledge and action in an age of epistemological turmoil: Notes on the deconstruction of knowing.* Paper delivered at the Seventh World Congress of Comparative Education, Montreal, Quebec.

Woolgar, S. (1988). *Science: The very idea.* London: Tavistock Publications.

Yahuda, M. B. (1978). *China's role in world affairs.* Great Britain: Redwood Burn Limited, Trowbridge & Esher.

Yahuda, M. B. (1983). *Towards the end of isolationism: China's foreign policy after Mao.* London: The Macmillian Press Ltd.

Yahuda, M. B. (1996). *The international politics of the Asia-Pacific, 1945-1995.* London: Routledge.

Zaghloul, M., & Altbach, P. (Eds.). (1993). *Higher education in international perspective: Toward the 21st century.* New York: Advent Books, Inc.

Zaghloul, M., & Altbach, P. (Eds). (1996). *Higher education in international perspective: Critical issues.* New York: Garland Publishing, Inc.

Zhi, X. (1989, January 23-29). Campus incident in Nanjing. *Beijing Review*, 7-9.

Appendixes A-N

Appendix A

A Sketch of Sino-African Relations in Historical Perspective

The founding of the People's Republic of China (1949) was the chosen point of entry for this investigation. In this century, contacts between China and Africa were rare before then. Historical connections, however, go back many centuries. The following sketch highlight these connections to provide some sense of the nature and depth of the historical linkages underlying the modern context of this investigation.

The First Encounters: Contact by Trade

While there is no official agreement as to the exact dates, modern Chinese scholars contend that Sino-African relations began even long before the Ming expeditions of the 15th century. These scholars maintain that China first traded with Sudan and Ethiopia in the Han dynasty (202 BC to AD 220) (Duyvendak, 1949; Filesi, 1962; Snow, 1988). Rulers of this time exchanged gold and silks for African luxury products such as ivory, rhinoceros horn, tortoise shell, frankincense, ambergris, pearls, rare stones, incense, copper, and camphor. Such goods were probably carried by Arab and Persian vessels which imported East African products to the southern ports of China and while there picked up Chinese exports to market in Africa. Thus, through merchant trafficking a link was forged between South China and East Africa (Duyvendak, 1949; Filesi, 1962; Snow, 1988).

African scholars tend to question such claims. While they agree that during these centuries African goods found their way to China and vice versa, they note that this evidence does not necessarily indicate that Chinese and African people actually came together (Fage, 1978; Portères & Barrau, 1981; Sheriff, 1981; Snow, 1988; Vérin, 1981). While trade to traditional Chinese sensibilities may have signified the existence of favourable relations, trade without the human dimension leaves many African scholars sceptical of such encounters (Snow, 1988). Thus, African historians are inclined to offer later dates of initial contact. Scholars account for these differing perspectives.

For more than two thousand years, in spite of foreign invasion and domestic revolution, China has remained the same recognizable political and cultural unit. Successive dynasties have maintained unbroken encyclopaedic records which detail evidence of China's past trading with foreign peoples. Such evidence rationalizes China's relationship with the same people today, and the older the historical evidence the stronger the modern connection is considered to be (Snow, 1988).

The African continent, on the other hand, does not have such unbroken encyclopaedic records. Ki-Zerbo (1981) states that for Africa the question of sources is a difficult one. He explains that African nations are faced with the task of reconstructing their history from written documents, archaeology, and oral tradition. While these three sources enable scholars to elaborate on and reinterpret history, they may not constitute unbroken encyclopaedic records.

In addition to the question of sources, M'Bow (1981: xvii) points out that scholars now widely recognize that the various civilizations of the African continent, for all their differing peoples, languages, and cultures, represent the historical offshoots of societies united by bonds centuries old. Unlike the bonds that unite China, however, the continent has not had the same recognizable political and cultural continuity (Fage, 1978; Snow, 1988). Since European colonial rule arbitrarily divided the continent just a century ago, African scholars tend to focus more on clarifying this recent past rather than exploring a possible encounter with other foreigners long before the Europeans came. For now, long distance trade through middlemen with a country an ocean away has limited meaning (Snow, 1988). Thus, there is no official agreement as to the exact date and nature of Sino-African relations, as sources of knowledge are scarce and historical accounts and perspectives vary (Duyvendak, 1949; Filesi, 1972; Snow, 1988).

Maritime Nations

Scholars write with more consensus about the historical connections between China and Africa beginning in the 15th Century. At that time, East African coastal regions were reaching the height of their prosperity as local merchants set sail on the Western Indian ocean seeking markets for their ivory in India and China (Hutchison, 1975; Fage, 1978; Sheriff, 1981). While African ships headed out, their own harbours welcomed a host of foreign fleets, many of which came from China. China, for the first time in history, had become a maritime nation and actually promoted overseas trade. The Emperor's Grand Eunuchs, entrusted with great fleets of over sixty ships and 40,000 soldiers, were ordered to inform the "barbarian" peoples of the "transforming power" of the "imperial virtue." In return, they were directed to receive tributes in acknowledgment of the Emperor's distant authority (Filesi, 1962; Hutchison, 1975: 11).[1]

The "most sensational salutation" tribute was a Bengal giraffe (Italiaander, 1961, as cited in Filesi, 1972:30).[2] This "gentle animal," with a "strange and marvellous structure," a "graceful walk," and "musical voice," was accorded almost divine honours. Poets celebrated, artists painted, and philosophers extolled the beauties and virtues of this wondrous animal (Duyvendak 1949; Filesi, 1972: 30). The African giraffe, as it strode into the Emperor's Court, became the emblem of perfection: perfect virtue, perfect government, and perfect harmony in the Chinese Empire and in the universe (Duyvendak, 1949; Filesi, 1972).

Yet while the giraffe symbolized perfection, it also symbolized dissonance. According to Italiaander, the Chinese seemed no longer satisfied with home and conceived the desire to know the mysterious territory of East Africa from where so many marvellous envoys and animals came (Italiaander, 1961, cited by Filesi, 1972: 30). Duyvendak (1949: 32) even goes so far as to conclude, "It was the giraffe, therefore, that caused the Chinese to sail to Africa."

Many did set sail. Today evidence of these voyages remains not only in Chinese documentation but also along the coasts and offshore islands of the four modern republics of Somalia, Kenya, Tanzania, and Mozambique. Across these nations, Chinese porcelain and pottery bits can be found decorating the walls and roofs of houses, mosques, and tombs. Moreover, in Kenya and Tanzania, copper coins of the Tang and Song dynasties account for approximately 300-500 pieces of the pre- modern foreign currency discovered. (Snow, 1988; Vérin, 1981). While these Chinese merchants and travellers left their mark along the coasts of East Africa, they also made a mark in China returning with spices, ivory, as well as information about the regions from which such goods came. The African continent, its products, and people became a subject of great interest, and a succession of writers set out to satisfy this demand.

Chinese Fictional and Factual Accounts

Chinese travellers' reports, official records, literary accounts, and other portrayals which emerged from the maritime voyages to East African reveal that the Chinese held contrasting views of Africans and their cultures (Hickey & Wylie, 1993; Snow, 1988; Sullivan, 1994). These texts show that though the Chinese merchants found the practices of the local people rather unseemly, as most non-Chinese practices were considered to be, the picture is not consistent. These accounts, characterized by a "textual middle ground," tended to refrain from judgement and instead stressed description of trade practices, landscape, wildlife (Duyvendak, 1949; Filesi, 1972; Hickey & Wylie, 1993; Snow, 1988).[3]

The fiction of the Chinese merchants in Africa corresponded in tone and manner with the fiction about Africans in China at that time. By the ninth century, Africans came to China from Arab dominions and were employed in Tang dynasty households. The short stories of this period described Africans, as workers in royal households, in such terms as brave, strong, athletic, resourceful, heroic, mysterious, and magical (Filesi, 1972; Snow, 1988: 17-20). Over the next few centuries, these attitudes changed.

From fictional accounts to factual records, the tone shifted from awe to disdain. According to documents from the year 1119, African people were 'kept' by the wealthy class in Canton, many of whom were members of a flourishing Islamic trading colony. Though the Africans did not necessarily work for the Chinese, the local residents saw them daily and no longer considered them special. When the mass slave trade reached its peak in Canton, Africans were increasingly seen as displaced nomads, ill-adapted to their Chinese surroundings.[4] Referred to as "devil slaves," they were thought to lack family ties and were classified by terms used to denote male and female animals (Snow, 1988: 18-19). These beliefs were firmly established when China closed itself off to the world from the 15th to the 19th century.[5]

China's Closure

The last medieval contact between China and Africa was recorded during the Ming dynasty when, in 1441, the ruler of Egypt, Sayyid Ali, sent envoys to the Imperial court with a "tribute of mules, horses, and various products of their localities" (Hutchison, 1975: 11). The next recorded official contact between the two regions came over five hundred years later when the then ruler of Egypt, President Gamul Abdel Nasser, was introduced to the Premier of the People's Republic of China, Zhou Enlai, at the Bandung Conference of African and Asian nations, April 24-27, 1955, Indonesia. At Bandung, China stepped out of its long isolation, reentered the world stage, and initiated the modern links between China and Africa (Hutchison, 1975: 11).

Hereafter, Zhou Enlai would stress "centuries old links", a "common heritage," and the "shared past" of China and Africa. Given the exact nature of those contacts, some historians find this stress rather surprising, yet most concede that China was neither aggressive nor plundering. Primarily, all China ever sought from Africa was a symbolic gesture of acquiescence to the Chinese view of the world — and it was the Chinese view of the world that brought them back together again (Snow, 1988: 29-30).

Notes

[1] Snow (1988: 29-30) compares China's first contacts with Africa to those of the Europeans' who came over seventy years later. First, China did not come to conquer but rather to garner prestige and profit for the Emperor in Peking. Second, the Chinese would never have thought Africa worth conquering as they believed extensive contact with non-Chinese peoples was neither desirable nor necessary. Africa was observed from too lofty a point to consider interrupting the flow of life.

[2] This giraffe, received by the Emperor in the hall of Receptions, was presented by renowned explorer, Cheng Ho (sometimes translated as Zheng He). Studies of Rockhill, Hirth, Pelliot, and Duyvendak offer factual evidence of Cheng Ho's fleets to the African coasts. These expeditions prove not only China's nautical proficiency,

but also the scale of the relationships at that period between continents almost entirely separate from the influences of the Western European world (Filesi, 1972: 2).

³ Hickey and Wylie (1993: 36) state that the themes, images, and assumptions that have informed China's textual construction of Africa were undoubtedly influenced by narrative and symbolic convention. In this regard, they show a number of constructional and conceptual similarities to the later European case.

⁴ For nearly one and a half millennia (a thousand years from about the 4ᵗʰ to the 14ᵗʰ century), African peoples were brought to China mostly by professional Arab traders who established a 'distributing centre' in Canton in 300 AD. Due to the absence of records or statistics, the actual number of Africans enslaved in China is impossible to estimate (Filesi, 1972; Wilbur, 1943, as cited by Sono, 1993: 25-26).

⁵ During the Ming dynasty (1368 to 1644), the Mongols were driven out and native Chinese rule restored. Ship building stopped, and within decades virtually no Chinese ships were on the oceans. The final voyage in 1433 signalled the end of the great merchant days. Imperial officials, determined that such mistaken policies not be followed again, burned all documents related to the voyages. Historians pronounced against expansionism, and the increasing influence of Confucian conformism quelled interest in the world beyond the Middle Kingdom. China inverted and withdrew from the outside world from the 15ᵗʰ to the 19ᵗʰ century (Hutchison 1975: 11).

Appendix B

Letter Announcing China/Canada Scholars & Students Exchanges Program
November 25, 1996

中 华 人 民 共 和 国 驻 加 拿 大 大 使 馆 教 育 处

EDUCATION OFFICE

THE EMBASSY OF THE PEOPLE'S REPUBLIC OF CHINA

IN CANADA

253

80 Cobourg Street
OTTAWA ONT.,
CANADA K1N 8H1

TELEPHONE (613) 789-6312
FAX   (613) 789-0262

Ms. Sandra Gilespie

March 25 1997

RE: China/Canada Scholars Exchange Program

It is my great pleasure to inform you that the Selection Committee of CCSEP has recommended you as one of the award recipients for the 1997/98 China/Canada Scholars Exchange Program. All your documents will be forwarded to the China Scholarship Council(CSC) for final approval. The CSC is also responsible for your placement in the Chinese University you have chosen. Please be advised that the approval and placement process would take about two months. An Official admission notice can therefore be expected sometime in the summer.

Please fill in the formal application form enclosed in duplication(please make photocopy). Please be noted that you don't have to provide any photo and research proposal if you have submit them with the original application forms.

Meanwhile, I am sending you a copy of the Physical Examination Record for Foreigner. Please have your Physical Check-up done by a government-recognized hospital or health facility at your earliest convenience. Please send the application forms and the physical Examination Record back to me by April 20 1997.

May I wish you all the best and every success in your future endeavour.

Sincerely yours

Yang Xinyu
Second Secretary(Education)
Chinese Embassy

Appendix C

Letter of Recommendation from the China/Canada Scholars Exchange Program
March 25, 1997

中 华 人 民 共 和 国 驻 加 拿 大 大 使 馆 教 育 处

251

EDUCATION OFFICE
THE EMBASSY OF THE PEOPLE'S REPUBLIC OF CHINA
IN CANADA

80 Cobourg Street
OTTAWA. ONT,
CANADA K1N 8H1

TELEPHONE: (613) 783-63·2
FAX: (613) 763-0262

25 November 1996

Re: China/Canada Scholars & Students Exchange Program

Dear Sir/Madam

It is my pleasure to inform you of the China/Canada Scholars & Students Exchange Program for 1997/98 academic year. The program is a bilateral government-sponsored project to encourage Canadians to study and do research in China. The detailed description and the application form is enclosed for your reference.

If you are interested in the program, I would encourage you to apply for this program and submit your application form and the necessary documents by 1 February 1997. The Selection Committee is happy to consider your application.

With best wishes

Sincerely yours

Yang Xinyu(Ms)
Second Secretary(Education)
Chinese Embassy
E-mail: xyang@buildlink.com

Appendix D

Letter of Confirmation from China/Canada Scholars Exchange Program
July 28, 1997

中华人民共和国驻加拿大大使馆教育处
EDUCATION OFFICE
THE EMBASSY OF THE PEOPLE'S REPUBLIC OF CHINA
IN CANADA

80 Cobourg Street
OTTAWA ONT.
CANADA K1N 8H1

TELEPHONE   (613) 789-6312
FAX   (613) 789-0262

July 28 1997

Miss.Sandra Gillespie

Re: China/Canada Scholars Exchange Program

Dear Miss. Sandra Gillespie

It is my pleasure to inform you that you have been admitted by Tongji University under the China/Canada Scholars Exchange Program. You are required to register at the university between September 1 and September 3 1997. The scholarship will cover tuition fees; on-campus accommodation; medical care; textbooks; living allowance and international air fare. The duration of your scholarship is from Sept. 1997-Dec. 1997.

Enclosed please find:
1. Admission Notice
2. Visa Application for Foreigners Wishing to Study in China(JW102)"
3. Visa Application Form
4. Instructions for Visa Application
5. Enrollment Procedures and Requirement for Foreign Students for the Academic year 1997/98
6. Contact Address of Selected Chinese Universities

Please read the above documents carefully and sign on 1,2,3 and send me back the following for a student visa to China by 10 August 1997:

1. Your passport valid beyond six month
2. Visa Application for Foreigners Wishing to Study in China(2 copies, one original, one photocopy. The original one will be returned to you with your

passport)

3. Completed Visa application Form with a passport-size photo glued on it

4. A money order or certified cheque of CN$52 payable to the Chinese Embassy

Please bring the "Visa Application for Foreigners Wishing to Study in China(original)", "Admission Notice" to register at the university and apply for the residency permit in the local authority.

Since the Canadian Dept. Of Foreign Affairs and International Trade is responsible for providing the international air-tickets, please contact Mr. Ryan Lock at (613)9925726 or Fax:(613)9925965 for the travel arrangements. You are required to inform your host institution in China your travel arrangements.

Finally, I wish to extend my warm congratulations and hope that you will have an enjoyable stay in China.

Sincerely yours

Yang Xinyu,
Second Secretary(Education)
Chinese Embassy

Appendix E

Admission Notice
June 6, 1997

ADMISSION NOTICE

Dear Ms. **GILLESPIE, SANDRA KATHRYN LOYCE:**
(Register No.**97124006**)

We are pleased to inform you that, having examined your application materials, we have decided to enroll you to study in the program of **China's Educational Contribution to Africa** taught in **Chinese** from 9/Mon. **97**/yr. to 7/Mon.**98**/yr.

According to the Advice of China scholarship Council, your fees for studying in China will be covered by:

Full scholarship (including tuition, lodging, medical care, learning material and living allowance) ☒

Partial scholarship: Tuition ☐ lodging ☐ medical care ☐ learning material ☐

If you observe the laws and the decrees of China government as well as the rules and regulations of the university you attend, and also accept the Additional Conditions as follows, you can apply for the student visa (**X visa**) to the Chinese embassy or consulate in your country with this Admission Notice, Visa Application Form for Foreigners Wishing to Study in China(**Form JW201**),the original copies of your Physical Examination Record for International Traveler and your blood test reports. Please note that you must register, with these materials, at the Foreign Student Office of Tongji University within the period between 8/Mon.30/Day and 9/Mon.3/Day, **1997**. If you fail to register within time limit without the permission of the hosting institution, you will be regarded as giving up this admission.

Additional conditions:

☐ In view of your Chinese language proficiency, you have to study the Chinese in the Department of Chinese language at our university for _____ year.

☒ If your Physical Examination Record is not up to the standard or is invalid, you need to be re-examined at Shanghai Sanitation and Anti-epidemic Station: and if you are suffering from any diseases which affect you study in China, you should return home country promptly and you have to bear all expenses by your own.

☒ You are admitted only as a general scholar, so you can neither pursue any degree nor extend your study in China.

Applicant's signature   **Tongji University**

Date: ____/Mon____Day/____Yr.   Date:6/Mon4Day/1997Yr.

Note: 1 Be sure to enter China with X visa, otherwise you will be responsible for all the possible consequences.

2. Please prepare eight copies of the photo the same size as in the passport.

3. Please go through all the procedures of registration to our university and apply within thirty days of arrival for the residency permit to the local police authority otherwise you will be fined for overdue.

Appendix F
Intended Interview Consent Form for African Students[6]

INTERVIEW CONSENT FORM FOR AFRICAN STUDENTS
This study focuses on the educational exchanges between China and Africa.

PURPOSE:
The main purpose for this study is to generate new knowledge about an increasingly important dimension of international education: South-South exchanges. Specifically, this research aims to obtain information about the lives of African scholars in China. Questions in the interview pertain to biographical information, motives for participation, applicability of education, and possibilities of future mutual exchanges.

PROCEDURE:
Interviews will be held at a location and time convenient to you. The length of interview should not normally exceed one hour. With your approval, I will tape the interview. After the interview, I will prepare a written summary of what has been said. I will show you this written summary to ensure that I have properly understood and captured your meaning.

BENEFITS OF INVOLVEMENT:
I would like to emphasize that your voluntary and anonymous participation in this work is of great significance. Ideally, the findings of this research will contribute on both a practical and theoretical level to all concerned parties. The results of this study aim to highlight and further enhance the cooperation that China and many African nations have sustained for more than 40 years. Moreover, I hope this research will broaden the understanding in the West about the kinds and nature of mutual support between China and Africa.

RIGHTS OF PARTICIPANT:
Your participation in this study is entirely voluntary. You will not receive nor be denied any personal benefits due to participation or non participation in this study. Names of person shall not be released in any publication which may result from these conversations. You are free to ask questions at anytime, to clarify statements, and to refuse to answer any interview questions. You are free to withdraw your participation at any time, for any reason. All data will from this study will be coded by participant number rather than by name. Findings will be reported in a strictly

confidential manner.

Your signature indicates that you agree to be interviewed and have read and understood this form and its contents. Having read and understood the information above, I hereby give my informed consent to be interviewed in the study.

———————————————

———————————————

Participant's signature
Date

———————————————

———————————————

Please print: Family Name
Given Names

Gender: ———————————
Country of Citizenship: ———————————
Number of years in China:———————————
Name of University: ———————————
Phone Number: ———————————

Appendix G
Intended Interview Consent Form for Chinese Participants

INTERVIEW CONSENT FORM FOR CHINESE PARTICIPANTS
This study focuses on the educational exchanges between China and Africa.

PURPOSE:
The main purpose for this study is to generate new knowledge about an increasingly important dimension of international education: South-South exchanges. Specifically, this research aims to obtain information about the lives of African scholars in China. Questions in the interview pertain to biographical information, motives for participation, applicability of education, and possibilities of future mutual exchanges.

PROCEDURE:
Interviews will be held at a location and time convenient to you. The length of interview should not normally exceed one hour. With your approval, I will tape the interview. After the interview, I will prepare a written summary of what has been said. I will show you this written summary to ensure that I have properly understood and captured your meaning.

BENEFITS OF INVOLVEMENT:
I would like to emphasize that your voluntary and anonymous participation in this work is of great significance. Ideally, the findings of this research will contribute on both a practical and theoretical level to all concerned parties. The results of this study aim to highlight and further enhance the cooperation that China and many African nations have sustained for more than 40 years. Moreover, I hope this research will broaden the understanding in the West about the kinds and nature of mutual support between China and Africa.

RIGHTS OF PARTICIPANT:
Your participation in this study is entirely voluntary. You will not receive nor be denied any personal benefits due to participation or non participation in this study. Names of person shall not be released in any publication which may result from these conversations. You are free to ask questions at anytime, to clarify statements, and to refuse to answer any interview questions. You are free to withdraw your participation at any time, for any reason. All data will from this study will be coded by participant number rather than by name. Findings will be reported in a strictly

confidential manner.

Your signature indicates that you agree to be interviewed and have read and understood this form and its contents. Having read and understood the information above, I hereby give my informed consent to be interviewed in the study.

_____
_____

Participant's signature
Date

_____
_____

Please print: Family Name
Given Names

Name of University: _____
Phone Number: _____

Appendix H

Cover Letter and Questionnaire in English

Dear International Student:

My name is Sandra Gillespie and I am a doctoral candidate at the University of Toronto in Canada. I have come to Shanghai to conduct research for my dissertation. My dissertation seeks the knowledge, opinions, and experiences of African scholars in the People's Republic of China. Three sites have been chosen for this investigation: Tongji University, Shanghai Medical University, and Zhejiang Agricultural University. These cites represent three different academic disciplines: Engineering, Medicine, and Agriculture respectively. As a student at one of the institutes, your participation in this project would be greatly appreciated.

First, I request your assistance in completing the enclosed questionnaire. The questionnaire is designed to generate information about your educational experience and life in China. Please choose to complete the questionnaire in either French or English. For each question, fill in the blank or check [√] the box that best indicates your response. Additional instructions are given where necessary. Please answer all eight sections of the questionnaire which will require approximately 30 minutes of your time. Once finished, please return the questionnaire in the envelope provided.

Second, I am seeking volunteers for one hour individual interviews. These interviews will supplement the findings generated by the questionnaire. The interviews will be held at a time and place convenient to you. Please indicate your willingness to be interviewed by contacting me at the address below.

Allow me to emphasize two points. First, your participation is entirely *voluntary*. Second, if you do freely choose to participate, your identity will remain strictly *confidential*. Please do not put your name on the questionnaire as all responses will be treated *anonymously*. The findings from the questionnaire and interviews will be presented in such a way that no individual respondent will be identifiable.

Your contributions to this study are essential. Should you have any questions, please feel free to contact me at Tongji university.

Thank you very much. I look forward to learning from you.

Sandra Gillespie

September 1997
Tongji University
Shanghai, China
200092

Questionnaire

SECTION 1: STUDENT PROFILE

Biographical Information
1. Nationality:_____
2. Sex: Male ☐ Female ☐   3. Age:_____
4. Mother Tongue(s):_____
5. Other languages you speak:

_____

6. Religion:_____
7. Marital Status: Single ☐ Married ☐   Other ☐
If presently married, is your spouse in China? Yes ☐ No ☐
8. Do you have any children?   Yes ☐ No ☐
If yes, are your children in China? Yes ☐ No ☐

Family Background

1. Please indicate the highest level of formal education of your parents.
Please check √ one box for each parent.

	Father	Mother
No formal education	☐	☐
Elementary	☐	☐
Secondary	☐	☐
Post secondary	☐	☐
Bachelor's degree	☐	☐
Master's degree	☐	☐
Doctoral degree	☐	☐
Other (specify)_____	☐	☐

2. Occupation of Parents: Father:_____
                          Mother:_____
3. How many siblings do you have?_____
4. What birth position do you occupy in your family? (1st, 2nd, 3rd born
etc.)_____
5. Where did you grow up?
          ☐ in the countryside   ☐ in a small city
          ☐ in a moderate size city   ☐ in a large city

Cultural Background

1. While growing up, how much personal exposure did you have to other cultures?

Please check √ one box for each category.

	1 Very great amount	2 Great amount	3 Moderate amount	4 Small amount	5 None
Other African cultures	☐	☐	☐	☐	☐
Asian	☐	☐	☐	☐	☐
Middle Eastern	☐	☐	☐	☐	☐
European	☐	☐	☐	☐	☐
North American	☐	☐	☐	☐	☐
South American/ Caribbean	☐	☐	☐	☐	☐
Other (specify)_____	☐	☐		☐	☐

2. Before coming to China, had you travelled outside of your country?
Yes ☐ No ☐
If yes, please state the three countries in which you stayed the longest.

	Country	Length of stay
1	_____	_____
2	_____	_____
3	_____	_____

Academic Background

1. What type of secondary school did you attend?
☐ Public
☐ Missionary
☐ Private
☐ Foreign (outside your country)
☐ Other(specify)_____

2. What kind of school was it?
☐ Boarding School
☐ Day School

3. In your secondary school, what was the main language(s) of instruction?_____

4. Were you the first child in your family to go to secondary school?
Yes ☐ No ☐

5. Are you the first child in your family to receive higher education?
Yes ☐ No ☐

6. Did you attend a university before coming to China?   Yes ☐ No ☐
    If yes, please indicate
      1. in which country(ies):_____
      2. which degree(s) you held before coming to China:_____
_____

7. Before leaving your home country, to what extent were you informed about your academic program and living conditions in China? Please check ☐ the box that best indicates your response.

Very well informed	Adequately informed	Fairly well informed	Not Adequately informed	Not informed at all
☐	☐	☐	☐	☐

8. From where/whom did you receive most of your information about China?_____

Current Education in China

1. Faculty:_____
2. Major/Specialization: _____
3. Level of study: Bachelor's ☐   Master's ☐   Doctoral ☐   Other ☐(specify) _____
4. How many years is your program?_____
5. What year are you in?_____
6. How many years have you been in China?_____

SECTION 2: MOTIVATION

1. At the time of your decision to leave your country to study in China, how important were the following factors? Please check [√] the box that best indicates the importance of each factor in motivating you to study in China.

	1 Very important	2 Important	3 Not applicable	4 Unimportant	5 Very Unimportant

Financial Motivations:

1. I obtained a scholarship. ☐ ☐ ☐ ☐ ☐
2. I received more financial aid by going to China than by studying at home. ☐ ☐ ☐ ☐ ☐
3. My family promised me assistance if I studied abroad. ☐ ☐ ☐ ☐ ☐

Academic Motivations:

4. In my country, I feared I would not be accepted into a university. ☐ ☐ ☐ ☐ ☐
5. In China, facilities in my field of study were better than in my country. ☐ ☐ ☐ ☐ ☐
6. In China, there were courses and facilities not available in my country. ☐ ☐ ☐ ☐ ☐

Employment Motivations:

7. In my country, a degree from China is worth more than a local degree. ☐ ☐ ☐ ☐ ☐

	1	2	3	4	5
	Very	Important	Not	Unimportant	Very
	important		applicable		

Unimportant

Employment Motivations:

8. I felt I could get a good job at home with a degree from China.

| ☐ | ☐ | ☐ | ☐ | ☐ |

9. I felt I could get a good job abroad with a degree from China.

| ☐ | ☐ | ☐ | ☐ | ☐ |

Personal Motivations:

10. Relatives and friends advised me to study abroad.

| ☐ | ☐ | ☐ | ☐ | ☐ |

11. Educational authorities advised me to study abroad.

| ☐ | ☐ | ☐ | ☐ | ☐ |

12. I wanted a chance to see the world.

| ☐ | ☐ | ☐ | ☐ | ☐ |

13. I wanted to get to know Chinese people and their customs.

| ☐ | ☐ | ☐ | ☐ | ☐ |

2. Please describe other factors that influenced you to study in China.

_____

_____

_____

_____

_____

_____

_____

_____

SECTION 3: ISSUES

1. Foreign students face many challenges. To what extent do you agree or disagree with each of the following statements? Please check √ the box that best indicates your response.

	1 Strongly agree	2 Agree	3 Neither agree nor disagree	4 Disagree	5 Strongly disagree
1. I enjoy living on campus.	☐	☐	☐	☐	☐
2. My Chinese classmates are friendly and helpful.	☐	☐	☐	☐	☐
3. I am often homesick.	☐	☐	☐	☐	☐
4. I am able to enjoy contact with fellow foreign students.	☐	☐	☐	☐	☐
5. I am able to enjoy close personal relationships with Chinese men.	☐	☐	☐	☐	☐
6. I am able to enjoy close personal relationships with Chinese women.	☐	☐	☐	☐	☐
7. I am concerned about racial discrimination.	☐	☐	☐	☐	☐
8. I suffer from occasional depression.	☐	☐	☐	☐	☐
9. I am often lonely in China.	☐	☐	☐	☐	☐
10. I feel highly motivated to do well in school.	☐	☐	☐	☐	☐
11. I feel I had adequate educational preparation at home.	☐	☐	☐	☐	☐
12. The attitude of the local people towards my country is favourable.	☐	☐	☐	☐	☐
13. My behaviour is often misunderstood by the local people.	☐	☐	☐	☐	☐
14. The faculty has a reasonable knowledge of my country.	☐	☐	☐	☐	☐

2. Do you think the previous list presents a complete picture of the most important issues facing foreign students in China today? Yes ☐ No ☐
If no, what do you think is missing?

_____
_____
_____
_____
_____
_____
_____

SECTION 4: SOCIAL CONTACT

1. When you are in the company of others, are they mainly:
☐ other Africans
☐ other foreigners
☐ people from China
☐ other (specify)_____
2. Do you have a girlfriend or boyfriend in China? Yes☐ No☐
If yes, is he/she: ☐ people from your own country
                 ☐ a person from your own country.
                 ☐ another African
                 ☐ another foreigner
                 ☐ a person from China
                 ☐ other (specify)_____

3. Please indicate the frequency with which the following statements apply to you.

	1 Daily	2 Weekly	3 Monthly	4 Yearly	5 Never
1. I have social contact with Chinese people.	☐	☐	☐	☐	☐
2. I have social contact with Chinese families.	☐	☐	☐	☐	☐
3. I engage in athletic activities with Chinese people.	☐	☐	☐	☐	☐
4. I watch Chinese television programs.	☐	☐	☐	☐	☐
5. I read Chinese newspapers and magazines.	☐	☐	☐	☐	☐

6. I participate in Chinese

student organizations. ☐ ☐ ☐ ☐ ☐

SECTION 5: ACADEMIC EXPERIENCE

1. Please rate your academic experience in China on the following items.

	1 Excellent	2 Good	3 Fair	4 Poor	5 Not applicable
1. Orientation of international students	☐	☐	☐	☐	☐
2. Faculty assistance	☐	☐	☐	☐	☐
3. Quality of instruction	☐	☐	☐	☐	☐
4. Relevance of course content to your home country	☐	☐	☐	☐	☐
5. Program design and requirements	☐	☐	☐	☐	☐
6. Intellectual stimulation in general	☐	☐	☐	☐	☐
7. Relevance of course content to your future plans	☐	☐	☐	☐	☐
8. Level of involvement in student activities	☐	☐	☐	☐	☐
9. Overall satisfaction with the Chinese educational program	☐	☐	☐	☐	☐

2. Please indicate the extent to which you experienced ease or difficulty in the following academic areas.

	1 Very easy	2 Easy	3 Not easy but not difficult	4 Difficult	5 Very difficult
1. Understanding lectures	☐	☐	☐	☐	☐
2. Writing term papers and assignments	☐	☐	☐	☐	☐
3. Taking notes in class	☐	☐	☐	☐	☐
4. Selecting courses	☐	☐	☐	☐	☐
5. Communicating with school authorities	☐	☐	☐	☐	☐
6. Using the library	☐	☐	☐	☐	☐
7. Establishing rapport with professors	☐	☐	☐	☐	☐
8. Speaking in front of					

others ☐ ☐ ☐ ☐ ☐
9. Getting academic advice ☐ ☐ ☐ ☐ ☐

	1	2	3	4	5
	Very easy	Easy	Not easy but not difficult	Difficult	Very difficult
10. Taking examinations	☐	☐	☐	☐	☐
11. Completing course work	☐	☐	☐	☐	☐
12. Working in cooperation with classmates	☐	☐	☐	☐	☐
13. Other (please explain)					

SECTION 6: CHINESE LANGUAGE AND PROGRESS

1. Did you receive any language training before coming to China?
Yes☐ No☐
If yes, specify length:_____
2. Did you receive any language training in China? Yes☐ No☐
If yes, specify length:_____
3. How would you rate your proficiency in the Chinese language?

	1	2	3	4	5
	Fluent	Advanced	Intermediate	Fair	Poor
Speaking	☐	☐	☐	☐	☐
Reading	☐	☐	☐	☐	☐
Writing	☐	☐	☐	☐	☐
Listening	☐	☐	☐	☐	☐
Overall ability	☐	☐	☐	☐	☐

	1	2	3	4	5
	Very great	Great	Moderate	Little	Not at all
4. To what extent are you satisfied with your academic progress?	☐	☐	☐	☐	☐
5. To what extent do you think your proficiency in the Chinese language is related to your academic progress?	☐	☐	☐	☐	☐

SECTION 7: SOURCES OF FINANCIAL SUPPORT

1. What are your sources of financial support? Please indicate
your sources of funding and the approximate percentage of
support from each source.
Check √ all applicable Approximate percentage
☐ Government of my country   _____%
☐ Government of China   _____%
☐ Foreign or international agency   _____%
☐ Family and/or relatives   _____%
☐ Personal Finances   _____%
☐ Other (specify) _____   _____%
                        _Percentages should add up to 100   %.
2. Are your financial resources sufficient ? Yes☐ No☐   If no, please
explain. _____
_____
_____

SECTION 8: FUTURE PLANS

1. What are your plans after you complete your current academic program ?
        ☐ to continue to study in China
        ☐ to study in another country
        (specify)_____
        ☐ to seek employment in my own country
        ☐ to seek employment in another country
        (specify)_____
        ☐ to begin prearranged employment
        ☐ to return to my former employment
        ☐ undecided
        ☐ other (specify) _____

If you have any additional information you wish to provide please do so in the space below or attach a separate page.

Please return the questionnaire in the envelope provided. I look forward to hearing from you about the interviews.

Thank you very much for completing this questionnaire.

Sandra Gillespie
University of Toronto
Canada

Appendix I

Cover Letter and Questionnaire in French

Cher Etudiant Internationaux:

Je m'appelle Sandra Gillespie et je suis candidate au doctorat à l'Université de Toronto au Canada. Je suis venu à Shanghai pour faire une enquête pour ma thèse. Ma thèse est base sur les connaissances, les points de vue et les expérience des étudiants africains en République Populaire de Chine. Trois sites ont été choisis pour la présente enquête. Il s'agit de l'Université Tongji de Shanghai, l'Université Médicale de Shanghai ainsi que l'Université d'Agriculture de Zhejiang. Ces universités représentent trois disciplines académiques différentes, à savoir: Génie, Médecine et Agriculture. Comme vous êtes étudiant à ces universités, votre participation serait grandement apprécie.

Premièrement, j'aimerais vous demander de compléter ce questionnaire en annexes. Ce questionnaire est conçu pour recueillir les information sur votre expérience académique et votre vie en Chine. Vous pouvez compléter ce questionnaire en français ou en anglais selon votre choix. Pour chaque question, écrivez dans les espaces vides ou cocher [√] la case qui indique mieux votre réponse. S'il vous plaît compléter toutes les huit sections de ce questionnaire. Cela devrait prendre a peu près 30 minutes de votre temps. Quand vous aurez termine de compléter ce questionnaire, s'il vous plaît renvoyez-le dans l'enveloppe annexée.

Deuxièmement, je recherche des volontaires pour des entrevues individuelles d'environ une heure de temps. Ces entrevues serviront à compléter les informations recueillies dans le questionnaire. Les entrevues se tiendront au temps et endroit de votre choix. S'il vous plaît indiquez votre intention de participer dans les entrevues en me contactant à l'adresse ci-bas.

Permettez-moi de souligner deux points. Premièrement, votre participation est entièrement *volontaire*. Deuxièmement, si vous choisissez de participer, votre identité restera strictement *confidentielle*. S'il vous plaît ne marquer pas votre nom sur le questionnaire car celui-ci sera traite *anonymement*.

Votre contribution à cette étude est essentielle. Si vous voudriez des éclaircissements, n'hésitez pas à me contactera l'Université de Tongji.

Je vous remercie beaucoup et j'attends impatiemment vos informations.

Sandra Gillespie

Septembre 1997
Université de Tongji
Shanghai, Chine
200092

Questionnaire
SECTION 1: PROFIL DE L'ETUDIANT

Informations bibliographiques

1.Nationalité: _____
2. Sexe: Masculin☐ Féminin☐ 3. Age:_____
4.Langue(s) maternelle(s): _____
5.Autres langues parlées:_____
6.Religion: _____
7.Etat civil: Célibataire☐ Marié(e)☐ Autre☐ Si vous êtes marié(e)s, votre conjoint(e) est-il (elle) en Chine? Oui☐ Non☐
8. Avez-vous des enfants? Oui☐ Non☐ Si oui, sont-ils en Chine? Oui☐ Non☐

Contexte familial

1. Veuillez indiquer le plus haut niveau d'éducation de vos parents. S'il vous plaît marquer [√] dans une case pour chaque parent.

	Père	Mère
Pas d'éducation formelle	☐	☐
Primaire	☐	☐
Secondaire	☐	☐
Collège	☐	☐
Bachelier	☐	☐
Maîtrise	☐	☐
Doctorat	☐	☐
Autre (expliquer)_____	☐	☐

2. Occupation des parents: Père: _____
                           Mère: _____
3. Combien de frères et soeurs avez-vous?_____
4. Quelle rang occupez-vous dans votre famille? (Exemple: 1er, 2ème, 3ème enfant, etc.)_____
5. Dans quel milieu avez-vous grandi?
☐à la campagne ☐dans une petite ville ☐dans une ville à dimensions moyennes ☐dans une grande ville

Contexte Culturel

1. Durant votre enfance, quel était le niveau de contact personnel avec les cultures étrangères?
S'il vous plaît marquer √ dans une case pour chaque catégorie.

	1 Très élevé	2 Élevé	3 Moyen	4 Bas	5 Nul
Autres cultures africaines	☐	☐	☐	☐	☐
Asiatiques	☐	☐	☐	☐	☐
Moyen-Orientales	☐	☐	☐	☐	☐
Européennes	☐	☐	☐	☐	☐
Nord-Américaines	☐	☐	☐	☐	☐
Sud-Américaines et Caraïbes	☐	☐	☐	☐	☐
Autres (expliquer)_____	☐	☐	☐	☐	☐

7. Avant de venir en Chine, aviez-vous voyagé en dehors de votre pays?
Oui☐ Non☐
Si oui, s'il vous plaît citer les trois pays dans lesquels vous êtes restés le plus longtemps.

	Pays	Durée de séjour
1	_____	_____
2	_____	_____
3	_____	_____

Préparation académique

1. Quel genre d'école secondaire avez-vous fréquenté?
  ☐ publique
  ☐ privée
  ☐ missionnaire
  ☐ étrangère (extérieure de votre pays)
  ☐ autre (expliquer)_____

2. Quelle sorte de régime était en place?
  ☐ Interne
  ☐ Externe

3. Dans votre école secondaire, quelle était la (les) langue(s) d'enseignement?

4. Etiez-vous le premier enfant de votre famille à faire l'école secondaire? Oui☐ Non☐

5. Etes-vous le premier enfant de votre famille à faire l'enseignement supérieur? Oui☐ Non☐

6. Etes-vous allé à l'université avant de venir en Chine? Oui☐ Non☐

Si oui, s'il vous plaît indiquer 1. dans quel pays: _____

                                       2. le diplôme que vous avez reçu:_____

7. Avant de quitter votre pays, combien étiez-vous informé à propos du programme académique et des conditions de vie en Chine? S'il vous plaît marquer ☐ dans la case qui indique mieux votre réponse.

Très bien informé	Adéquatement informé	Assez bien informé	Pas adéquatement informé	Pas du tout informé
☐	☐	☐	☐	☐

8. De quelle source avez-vous reçu le plus d'informations à propos de la Chine?_____

Education en cours

1. Faculté:_____

2. Spécialité:_____

3. Niveau d'étude: Bachelier☐ Maîtrise☐ Doctorat☐

Autre (expliquer)_____

4. Combien d'années faut-il pour terminer votre programme?_____

5. Dans quelle année êtes-vous? _____

6. Depuis combien de temps êtes-vous en Chine? _____

SECTION 2: MOTIVATION

1. Au moment de la prise de votre décision d'étudier en Chine, quelle était l'importance des facteurs suivants? S'il vous plaît marquer √ dans la case qui indique mieux l'importance de chaque facteur dans votre motivation pour étudier en Chine.

	1 Très important	2 Important	3 Neutre	4 Peu important	5 Pas du tout du tout important
Motivations financières:					
1. J'ai reçu une bourse d'étude.	☐	☐	☐	☐	☐
2. J'ai une meilleure aide financière en Chine qu'au pays natal.	☐☐	☐☐	☐☐	☐☐	☐☐
3. Ma famille m'a promis une assistance si j'allais étudier à l'étranger.	☐	☐	☐	☐	☐
Motivations académiques:					
4. Dans mon pays, je n'étais pas sûr d'être accepté à l'université.	☐	☐	☐	☐	☐
5. Les équipements académiques dans mon domaine d'étude étaient meilleurs en Chine que dans mon pays.	☐	☐	☐	☐	☐
6. En Chine, il y avait des cours qui n'étaient pas disponibles dans mon pays.	☐	☐	☐	☐	☐
Emploi:					
7. Dans mon pays, un diplôme de la Chine vaut plus qu'un diplôme local.	☐	☐	☐	☐	☐
8. J'ai pensé qu'avec un diplôme de la Chine, j'aurais un bon emploi chez-moi.	☐	☐	☐	☐	☐
9. J'ai pensé qu'avec un diplôme de la Chine, j'aurais un emploi à l'étranger.	☐	☐	☐	☐	☐

	1	2	3	4	5
	Très important	Important	Neutre	Peu important	Pas du tout du tout important

Motivations personnelles:

10. Mes amis et connaissances m'ont conseillé d'aller étudier à l'étranger. ☐ ☐ ☐ ☐ ☐

11. Les autorités en éducation m'ont conseillé d'aller étudier à l'étranger. ☐ ☐ ☐ ☐ ☐

12. Etudier en Chine constituait une chance de voir le monde. ☐ ☐ ☐ ☐ ☐

13. J'avais l'intention de mieux connaître le peuple ☐ ☐ ☐ ☐ ☐

2. S'il vous plaît décrire les autres facteurs qui vous ont poussé à étudier en Chine.

SECTION 3: ASPECTS DIVERS

1. Les étudiants étrangers font face à différents défis. A quel degré êtes-vous en accord ou en désaccord avec les affirmations suivantes? S'il vous plaît marquer √ dans la case qui indique mieux votre réponse.

	1 Très d'accord	2 D'accord	3 Neutre	4 Pas d'accord	5 Pas du tout d'accord
1. J'aime vivre au campus.	☐	☐	☐	☐	☐
2. Mes camarades de classe sont gentils et serviables envers moi.	☐	☐	☐	☐	☐
3. J'ai souvent envie de rentrer chez-moi.	☐	☐	☐	☐	☐
4. J'ai de bonne relations avec les autres étudiants étrangers .	☐	☐	☐	☐	☐
5. Je suis capable d'entretenir des relations personnelles avec des Chinois.	☐	☐	☐	☐	☐
6. Je suis capable d'entretenir des relations personnelles avec des Chinoises.	☐	☐	☐	☐	☐
7. La discrimination raciale m'inquiète.	☐	☐	☐	☐	☐
8. Je suis occasionnellement déprimé	☐	☐	☐	☐	☐
9. J'éprouve souvent la solitude.	☐	☐	☐	☐	☐
10. Je ressens une grande motivation de très bien réussir en classe.	☐	☐	☐	☐	☐
11. J'ai eu une préparation académique adéquate dans mon pays.	☐	☐	☐	☐	☐
12. L'attitude de la population locale à l'endroit de mon pays natal est favorable.	☐	☐	☐	☐	☐
13. Mon comportement est souvent mal compris pa la population locale.	☐	☐	☐	☐	☐

	1	2	3	4	5
	Très	D'accord	Neutre	Pas	Pas du
	d'accord			d'accord	tout d'accord
14. Les membres de ma Faculté ont une bonne connaissance de mon pays.	☐	☐	☐	☐	☐

2. D'après vous, la liste précédente présente-elle une image complète des aspects auxquels les étudiants étrangers en Chine sont confrontés aujourd'hui? Oui☐ Non☐ Si non, qu'est-ce qui manque, à votre avis?

SECTION 4: RELATIONS SOCIAUX

1. Quand vous êtes en compagnie d'autres personnes, sont-elles principalement:
   - ☐ des ressortissants de votre pays
   - ☐ d'autres Africains
   - ☐ d'autres étrangers
   - ☐ des Chinois
   - ☐ autre (expliquer) _____

2. Avez-vous une copine ou un copain en Chine? Oui☐ Non☐ Si oui, est-elle (il)
   - ☐ une personne de votre pays
   - ☐ un(e) Africain(e)
   - ☐ une personne étrangère
   - ☐ une personne Chinoise
   - ☐ autre (expliquer)_____

3. S'il vous plaît, indiquer la fréquence à laquelle les affirmations suivantes sont applicables à votre cas.

	1 Quotidien	2 Hebdomadaire	3 Mensuel	4 Annuel	5 Jamais
1. J'ai un contact social avec les Chinois.	☐	☐	☐	☐	☐
2. J'ai un contact social avec des familles chinoises.	☐	☐	☐	☐	☐
3. Je participe dans des activités sportives avec les Chinois.	☐	☐	☐	☐	☐
4. Je suis les programmes de la télévision chinoise.	☐	☐	☐	☐	☐
5. Je lis les journaux et magasines chinois.	☐	☐	☐	☐	☐
6. Je participe dans les organisations estudiantines chinoises.	☐	☐	☐	☐	☐

Continued on next page

SECTION 5: EXPÉRIENCES ACADÉMIQUES

1. S'il vous plaît évaluer votre expérience académique en Chine pour les détails suivants.

	1 Excellent	2 Bon	3 Passable	4 Médiocre	5 Sans objet
1. Orientation des étudi--ants internationaux	☐	☐	☐	☐	☐
2. Assistance des membres de la faculté	☐	☐	☐	☐	☐
3. Qualité de l'enseignement	☐	☐	☐	☐	☐
4. Pertinence du contenu des cours pour votre pays	☐	☐	☐	☐	☐
5. Plan et exigence des cours	☐	☐	☐	☐	☐
6. Stimulation intellectuelle	☐	☐	☐	☐	☐
7. Pertinence du contenu des cours pour vos plans d'avenir	☐	☐	☐	☐	☐
8. Niveau de participation dans les activités estudiantines	☐	☐	☐	☐	☐
9. Satisfaction du programme d'éducation chinois en général	☐	☐	☐	☐	☐

2. S'il vous plaît indiquer le niveau d'aisance ou de difficulté dans les domaines académiques suivants.

	1 Très facile	2 Facile	3 Ni facile Ni difficile	4 Difficile	5 Très difficile
1. Comprendre les exposés	☐	☐	☐	☐	☐
2. Ecrire les essais et faire les travaux à domicile	☐	☐	☐	☐	☐
3. Prendre les notes en classe	☐	☐	☐	☐	☐
4. Faire le choix des cours	☐	☐	☐	☐	☐
5. Composer avec les autorités de l'université	☐	☐	☐	☐	☐
6. Utiliser la bibliothèque	☐	☐	☐	☐	☐
7. Etablir des liens avec les professeurs	☐	☐	☐	☐	☐
8. Parler en face d'autres étudiants (en classe)	☐	☐	☐	☐	☐

	1	2	3	4	5
	Très facile	Facile	Ni facile Ni difficile	Difficile	Tres difficile
9. Recevoir des conseils académiques	☐	☐	☐	☐	☐
10. Passer les examens	☐	☐	☐	☐	☐
11. Compléter les travaux de cours	☐	☐	☐	☐	☐
12. Travailler en groupe avec les camarades de classes	☐	☐	☐	☐	☐

13. Autre (expliquer)_____

SECTION 6: LANGUE CHINOISE ET PROGRÈS

1. Avez-vous étudié le chinois avant de venir en Chine? Oui☐ Non☐ Si oui, indiquer la durée:_____
2. Avez-vous reçu une formation en langue chinoise en Chine? Oui☐ Non☐ Si oui, indiquer la durée:_____
3. Comment évaluez-vous vos compétences dans l'utilisation de la langue chinoise?

	1	2	3	4	5
	Courant	Avancé	Intermédiaire	Passable	Pauvre
Parler	☐	☐	☐	☐	☐
Lire	☐	☐	☐	☐	☐
Ecrire	☐	☐	☐	☐	☐
Comprendre à l'audition	☐	☐	☐	☐	☐
Compétence générale	☐	☐	☐	☐	☐

	1	2	3	4	5
	Très haut	Haut	Moyen	Bas	Très bas
4. A quel degré êtes-vous satisfait de votre évolution académique?	☐	☐	☐	☐	☐
5. A quel degré pensez-vous que votre compétence en chinois est lié à votre évolution académique?	☐	☐	☐	☐	☐

Continued on next page

SECTION 7: SOURCES DE FINANCEMENT

1. Quelles sont vos sources de financement? S'il vous plaît indiquer le pourcentage approximatif pour chaque source.

S'il vous plaît marquer [√] là où applicable Pourcentage approximatif
- ☐ Le gouvernement de mon pays d'origine _____%
- ☐ Le gouvernement chinois _____%
- ☐ Une agence internationale ou étrangère _____%
- ☐ Ma famille et /ou mes parentés _____%
- ☐ Financement personnel _____%
- ☐ Autres (expliquer)_____ _____%

☐ Le total doit être   100   %

2. Vos ressources financières sont-elles suffisantes?   Oui☐   Non☐   Si non, s'il vous plaît expliquer.

SECTION 8: PLANS D'AVENIR

1. Qu' est-ce que vous envisagez de faire à la fin de votre programme académique?
- ☐ continuer mes études en Chine
- ☐ continuer mes études dans un autre pays (expliquer)_____

chercher du travail dans mon pays
- ☐ chercher du travail dans un autre pays (expliquer)_____

- ☐ commencer le travail prévu pour moi dans mon pays
- ☐ retourner à mon ancien emploi
- ☐ indécis
- ☐ autre (expliquer)_____

Si vous avez des informations supplémentaires à donner, vous pouvez le faire dans l'espace suivant ou attacher une page supplémentaire.

S'il vous plaît retourner le questionnaire dans l'enveloppe annexée. J'attends votre réponse à propos des entrevues.

Merci infiniment de compléter ce questionnaire.
Sandra Gillespie
Université de Toronto
Canada

Appendix J

Intended Interview Schedule for African Students

Interview Schedule for African Students

BIOGRAPHICAL INFORMATION

Age:_____
Gender:____
Home Country: _____
Number of Years in China:_____
Field: Engineering, Medicine, Agriculture
Level: Bachelor's, Master's, Doctoral

EQUITY
Preparation
Orientation
Motivation

AUTONOMY
Level of prior information and knowledge

SOLIDARITY
Educational Priorities
China's response to those to those priorities
Relevance and applicability to home conditions

PARTICIPATION
Acquiring and sharing knowledge: lectures, class participation, involvement, campus life

OTHER
South-South Relations
Future Plans/Anticipations

Date:_____

Appendix K

Intended Interview Schedule for Chinese Participants

Interview Schedule for Chinese Participants

BIOGRAPHICAL INFORMATION

Position:_____
Institute: Engineering, Medicine, Agriculture

PROFILE OF PROGRAM AND STUDENTS
History and development of program
Formal characteristics of the program: number of students, countries, programs
Scholarship information

EQUITY
Preparation
Recruitment and admissions methods and philosophy
Application and selection process
Motivation: Aims and nature of China's educational aid for Africa

AUTONOMY
Level of prior information and knowledge

SOLIDARITY
Educational Priorities
China's educational practices in response to African development priorities
Relevance and applicability to conditions in African nations

PARTICIPATION
Acquiring and sharing knowledge: lectures, class participation, involvement, campus life, accommodation of the presence of foreign students

OTHER
South-South Relations
Future Plans/Anticipations

Date:_____

Appendix L

Original and Translated Document from the Department of Foreign Affairs of the State Education Commission People's Republic of China Dated April 18, 1997

我国与非洲的教育交流与合作

我国与非洲的教育交流与合作基本上可分为四个阶段:

一. 建国初期至 1966 年

在新中国成立不久的 50 年代初期,我国即接收了来自非洲国家的留学生,. 由于这一时期,许多非洲国家尚未独立,我只接受了埃及、喀麦隆、肯尼亚和乌干达的十几名学生,而在其它方面交流甚少。

60 年代,非洲国家争取独立的斗争风起云涌,我国对非洲人民的民族独立和解放运动给予了坚决支持。周恩来总理访问亚非 14 国时提出的中国同非洲和阿拉伯国家相互关系的五项原则和中国对外援助的八项原则深得非洲国家赞同,使中非关系得到很大发展。自 1960 年 9 月开始,经由我亚非团结委员会、中非友协、全国总工会等单位联系,喀麦隆人民联盟、桑桑巴尔民族主义党、索马里民族联盟、加纳工会等党派和群众团体向我国成批派遣留学生。截止到 1966 年底,共有来自 14 个国家的 190 余名非洲留学生来我国学习。我国也曾向埃及、摩洛哥和阿尔及利亚国家派遣过留学生。

在这期间,其它教育交流活动也开始增加。 到 66 年底,我教育部门有 5 个代表团组访问了非洲 8 个国家,其中教育部长杨秀峰和副部长刘皑风分别于 1964 和 1966 年率团访问了埃及、阿尔及利亚、马里、几内亚、坦桑尼亚、摩洛哥、中非等国。埃及、摩洛哥、几内亚、坦桑尼亚和中非等国亦应我国邀请派团访问了我国。此外,我国还在 1955 年和 1965 年分别向埃及和马里派遣了汉语和数理化教师,向坦桑尼亚和索马里提供了教学器材等援助。

二. 1966 年至 1978 年

文化大革命开始后,我国曾一度停止对外教育交流活动。到 1970 年才逐步恢复互派代表团和对非交流。从这年起开始向刚果派汉语和数理化教师,此后又向突尼斯和阿尔及利亚派了汉语教师。1971 年,联合国恢复了我国的合法席位,与我建交的国家已达百余个。我国的国际威望迅速提高。在这样的外交形势下,要求与我交往或恢复交往的国家越来越多,特别是独立后与我建交的非洲国家。我国于 1973 年恢复接受外国留学生,至 1978 年,共接收了 25 个非洲国家的近 500 名学生。在这一时期,亚非国家的学生占主导成分。非洲各国来华访问的教育代表团开始增多,先后有苏丹、索马里、坦桑尼亚、阿尔及利亚、贝宁、扎伊尔、卢旺达、赞比亚、埃塞俄比亚、刚果和几内亚等国团组来访,其中贝宁、阿尔及利亚、刚果、坦桑尼亚和赞比亚国为教育部长。这时期,我国出访的团组较少,只有 4 个代表团组访问了埃及等 3 个国家。

三. 1979 年至 1989 年

自 1978 年末我国实行改革开放以来,随着国民经济的迅速发展和我国国际地位的提高,我国与世界各国在教育方面的交流活动也日益频繁。非洲国家将我视为第三世界国家的榜样,纷纷要求与我交流与合作。双边互派团组急剧增加。到 89 年底,我出访非洲国家的团组近 40 个,访问了非洲 20 余个国家。

亦率团访问了非洲. 非洲也有 20 多个国家的 27 个团组来访. 越来越多的非洲国家希望我们为其培养建设人才, 来华留学生的规模和数量大大增加. 这时期, 向我国派遣留学生的非洲国家已从 70 年代的 25 个增加到 43 个, 人数也猛增到 2271 人. 我国也向埃及、摩洛哥、突尼斯、尼日利亚、坦桑尼亚等国派了少量的留学生.

在这一阶段, 根据我国对非援助的总体方针, 应非洲国家的要求, 我国开始持续向刚果、赞比亚、毛里塔尼亚、加蓬、毛里求斯、阿尔及利亚、埃及、突尼斯和马达加斯加等国派遣教师, 还向贝宁、马里、毛里塔尼亚、阿尔及利亚、卢旺达、布隆迪、乍得、刚果和毛里求斯等国赠送了小额教学用品.

四. 1990 年至 1996 年

我国的改革开放政策带来了经济的飞跃发展, 综合国力大大加强. 非洲国家将我视为 21 世纪最有希望的国家, 团组往来和互派留学生的数量不断增加, 特别加强了高层往来. 1996 年, 江泽民主席作为中国最高领导人首次访问非洲 5 国, 使中非关系进入了新的阶段. 教育领域的中非关系得到空前发展. 在短短的几年内, 国家教委主任朱开轩、副主任邹时炎、柳斌和张天保先后访问了非洲 15 国. 出访的各类团组达 40 余个, 来访的团组近 30 个, 其中部级团组就有 10 个.

为了适应国际形势的发展、变革和政治、经济与发展环境的需要, 我国调整了接受来华留学生的政策. 对非洲国家采取了 "高层次、短学制、高效益" 的方针, 缩减了本科生的人数. 在这期间共接受了 45 个国家的 1500 人. 与此同时, 我国也向非洲的 9 个国家派遣了近百名交换奖学金生.

除了传统的交流项目, 根据我国的外交政策和中央有关加强对第三世界国家、尤其是对非洲国家的智力援助的指示精神, 我与非洲国家的交流与合作无论从形式还是内容上都有了很大变化和发展, 从 1990 年起, 我开始与非洲国家开展教育合作项目, 旨在帮助非洲国家发展高等教育, 在当地培养本国所需的人才. 我们选择了受援国大学的薄弱学科或专业, 帮助其建立实验室并派教师任教, 有的还教汉语和中国文化. 这种合作项目受到非洲国家的普遍欢迎并取得很好成效. 6 年来, 我们与喀麦隆、马里、刚果、乍得、科特迪瓦、布隆迪、扎伊尔、塞内加尔、埃及、苏丹、坦桑尼亚、赞比亚、加纳、津巴布韦、肯尼亚、纳米比亚、毛里求斯、尼日利亚 18 个国家进行了 24 项项目, 已建立生物、计算机、物理、食品保鲜与加工、土木工程与测量和汉语等专业实验室, 派出教师 50 余人次.

智力援非项目的开展还带动了中-非大学之间的校际交流与合作. 到目前为止, 我国已有 10 所大学与非洲 16 个国家的 20 所大学建立了校际关系.

统计资料表明, 我国自 50 年代起至今, 已访问非洲的代表团达 90 余个, 接待的非洲国家代表团 80 多个; 已接受非洲留学生 4570 余名, 派赴非洲工作的教师 400 余人, 为非洲的 25 个国家提供过不同程度的教育援助. 这些交流与合作密切了我国与非洲大陆的关系, 加强了了解和友谊, 有助于我整体外交工作的顺利开展.

国家教委外事司供稿

Translated Summary of Document[1]
Department of Foreign Affairs
State Education Commission
People's Republic of China
April 18, 1997

The Educational Exchanges and Cooperation between China and Africa

In general, the educational exchanges and cooperation between China and Africa can be divided into the following four periods: 1949 to 1966, 1966 to 1978, 1979 to 1989, and 1990 to 1996.

First Period: 1949-1966

Approximately one dozen exchange students from Egypt, Cameroon, Kenya, and Uganda first came to the People's Republic of China shortly after its establishment in 1949. In 1955, China sent language teachers to Egypt and mathematics, physics, and chemistry teachers to Mali. Exchanges in other fields were limited because at this time many African nations, with the firm support of China, were struggling for independence.

Educational exchanges began to flourish in September 1960 when African students began to come to China. These students were sent by African political parties and mass organizations, such as the People's Coalition of Cameroon, the Nationalism Party of Zimbabwe, the National Alliance of Somalia, and the Trade Union of Ghana, in coordination with China's Asia-Africa Unity Committee, Sino-Africa Friendship Association, and the General National Trade Union.

Chinese delegations also made their way to Africa. Between 1963-65, Prime Minister Zhou Enlai visited 14 Asian and African countries proposing two agreements regarding the relationships between China and Africa. The first agreement, 'Five Principles on the Development of the Relationship between China and Africa' and the second agreement, 'Eight Principles on China's Support to Foreign Countries' were well received and enhanced the development of Sino-African relations. By the end of 1966, five Chinese educational delegations, two headed by the Minister and Vice Minister of Education, visited eight African countries including Egypt, Algeria, Mali, Guinea, Tanzania, Morocco, and Central Africa.[2] In 1966, China provided Tanzania and Somalia with teaching facilities. At the invitation of the Chinese government, delegations from Egypt, Morocco, Guinea, Tanzania, and Central Africa visited China. By the end of 1966, a few Chinese students were sent to Egypt, Morocco, and Algeria, and 190 African students from 14 countries were studying in China.

Second Period: 1966 to 1978

During the Cultural Revolution, all educational exchanges ceased. In 1970, China resumed the exchanges by sending language, mathematics, physics and chemistry teachers to the Congo and later sent language teachers to Tunisia and Algeria. In 1971, China regained its seat in the United Nations and established diplomatic relations with more than one hundred countries. As China's international prominence rose, many African nations, especially those which had association with China after their independence, sought to restore relations. From 1973, China resumed admitting foreign students, and by 1978 China had received almost 500 students from 25 African countries. In this period, Asian and African students constituted the majority of the foreign students in China.

In the meantime, China decreased the number of delegations going to Africa (only four delegations were sent to three countries including Egypt) while African nations increased the number of educational delegations visiting China. African delegations came from Sudan, Somalia, Tanzania, Algeria, Benin, Zaire, Rwanda, Zambia, Ethiopia, the Congo, and Guinea. Of the delegations listed above, those from Benin, Algeria, the Congo, Tanzania, and Zambia were headed by their Ministers of Education.

Third Period: 1979 to 1989

As a result of the Reform and Open Policy introduced in 1978, China's national economy and international status soared. Emerging as a model of the Third World, many African nations looked to China to train their personnel for economic construction. Accordingly, the number of educational exchanges and visiting missions from each side increased sharply. By the end of 1989, nearly 40 Chinese delegations visited 20 African nations, and 27 delegations from more than 20 African countries visited China.[1]

The number of African countries sending students to China increased from 25 in the 1970s to 43 in late 1980s. During this period, the total number of African students amounted to 2,271. Meanwhile, China sent a few students to study in Egypt, Morocco, Tunisia, Nigeria, and Tanzania. At this time, based on a general policy of support to Africa and at the request of individual nations, China began its sustained detachment of teachers to Africa. The recipient countries included the Congo, Zambia, Mauritania, Gabon, Mauritius, Algeria, Tunisia, and Madagascar. China also donated a modest amount of teaching equipment to Benin, Mali, Mauritania, Algeria, Rwanda, Burundi, Chad, the Congo, and Mauritius.

Fourth Period: 1990 to 1996

China's Reform and Open Policy brought about rapid economic development and national empowerment. As a result, many African countries began to view China as the most promising country in the twenty-first century. Accordingly, the contacts between the two sides, particularly in education, developed on an unprecedented scale. During this period, nearly 30 African delegations visited China, and more than 40 Chinese delegations visited over 15 African nations. Among these delegations were 10 ministerial level visits that included Zhu Kaisuan, the Director of the State Education Commission, and Zhou Shiyan, Liu Bin, and Zhang Tianbao, Vice Directors of the State Education Commission. The highlight came in 1996, when Chairman Jiang Zemin made his first visit to five African countries, signalling a new era of Sino-African relations.

In light of the emerging changes of the international situation, China's cooperation with Africa developed both in form and content. These developments came as a natural outgrowth of China's diplomatic policy and the Party's guidelines for providing intellectual support to Third World countries in general and African countries in particular. In 1990, China initiated a new policy of educational cooperation aimed at helping African nations develop their higher education systems and train local personnel. This new policy, described as "High Level, Short Term, High Benefit," shifted the focus from undergraduate to graduate education. In this period, China received 1,500 students from 45 African countries and sent out nearly 100 students to 9 African countries. In addition, China dispatched teaching staff and established laboratories in the universities of recipient countries. The laboratories specialize in different subjects, such as biology, computer, physics, food processing and freshness preservation, civil engineering construction, land survey, and the Chinese language.

In the past 6 years, China has carried out 24 cooperative programs in 18 African countries. The recipient countries include Cameroon, Mali, the Congo, Chad, Ivory Coast, Burundi, Zaire, Senegal, Egypt, the Sudan, Tanzania, Zambia, Ghana, Zimbabwe, Kenya, Namibia, Mauritius, and Nigeria. Furthermore, 10 Chinese universities have established school level associations with 20 universities in 16 African countries.

In all, statistics shows that since the 1950s, China has sent more than 90 delegations, dispatched more than 400 teachers, and provided various kinds of educational support to 25 African countries. In addition, China has hosted more than 80 delegations from African countries and admitted approximately 4,570 African exchange students. These exchanges have strengthened the relationship between China and the continent of Africa, promoted understanding and friendship, and aided the general development of our diplomacy.

Department of Foreign Affairs
State Education Commission
People's Republic of China
April 18, 1997

Notes

.¹ This document was obtained from China's State Education Commission (December 1997) and translated by Zhang Xiaoman in Shanghai (January 1998) and Pei Chao in Montreal (June 1998).

.² Among these delegations, one in 1964 was headed by the Minister of Education, Yang Xiufeng, and another in 1966 was headed by the Vice Minister of Education, Liu Kaifeng.

.³ Those who headed the delegations included Li Tieying, a member of the State Council and Director of State Education Commission; Peng Peiyun, Vice Minister of Education; and Teng Tengyi, Vice Director of the State Education Commission.

Appendix M

Notes

Unfortunately my copy of this memorandum was faint and askew. The tables were impossible to read. For this reason, I have chosen to remove the tables.

Appendix M

Appendix N

Letter from Master's and Doctoral Foreign Students
to the Director of the Foreign Students' Office
Tongi University, Shanghai
People's Republic of China

Notes

Unfortunately my copy of this letter was faint and askew. The type was impossible to read. For this reason, I have chosen to remove the tables.

Index

Printed and bound by CPI Group (UK) Ltd, Croydon, CR0 4YY

22/10/2024

01777603-0004